Producing Power Bucks
The Banks Farm Way

PRODUCING POWER BUCKS
The Banks Farm Way

How to Grow Bigger Whitetails in the Southeast
Through Food Plots and Supplemental Feeding

By Jeff Banks
And Duncan Dobie

Bucksnort Publishing, Ltd.
Marietta, Georgia
2002

Copyright © 2002 by:
Jeff Banks and Duncan Dobie

All Rights Reserved

Published by:
Bucksnort Publishing, Ltd.
P.O. Box 670794
Marietta, Georgia 30066

No part of this book may be reproduced in any form without
Written permission from Bucksnort Publishing, Ltd.
and Jeff Banks and Duncan Dobie.

Manufactured in the United States of America
Library of Congress Catalog Card Number: 2002108633
ISBN 0-939801-10-8

First Edition
Printed by Vaughan Printing Co.
Nashville, Tennessee

Dedication

I'd like to dedicate this book to my heavenly father Jesus Christ. Because of Him, we have been given this magnificent creation called Earth.

We are so blessed and grateful. It is in Him that we live and move and have our being.

I challenge each of you as well as myself to make everything in our world – our lives, our families, our relationships, our occupations and our recreation – a better place.

Also, to my beautiful and wonderful wife Michelle. Without your support and help, I could not have accomplished our goals on Banks Farm. I love you, I'm in love with you, and I'm blessed that you're my wife.

JLB III

Acknowledgements

Special thanks to **Georgia Ourdoor News** for allowing us to use the great cover photo taken by **Brad Bailey**. The photo first appeared on the December, 2001 cover of GON. Also, thanks for the use of the photos on pages 39 and 209.

Thanks to **Charles Hendrix** for allowing us to use several of his excellent trail camera photos.

Special thanks to **Allen Hansen** for his great illustration of the "Horn Donkey."

Special thanks to **Gordon Whittington**, editor of **North American Whitetail** magazine, for reading the rough manuscript of this book and giving us some keen insight and advice on how to make it better.

Special thanks to **Pat McDevitt** for his help in editing this book.

Producing Power Bucks
The Banks Farm Way

Table of Contents

Chapter		Page
	Introduction	8
1	A Legend is Born	16
2	How We Got Started on the Banks Farm	24
3	Nutrition, Nutrition, Nutrition	48
4	High-Protein Food Plots	66
5	Supplemental Feeding	88
6	Mineral Supplements	96
7	Putting It All Together – Low Cost Planting/Supplemental Feeding	102
8	Other Management Considerations	112
9	Using Trail Cameras and Video Cameras	132
10	Goals for Your Hunting Club	144
11	Seeing Big Results – Sweat Equity and Making the Dream a Reality	172
12	The Legacy of the Horn Donkey	186
13	Banks Farm Checklist – Valuable Resources for Trophy Management	204

After 12 years of intensive trophy management, the Banks Farm in Morgan County, Georgia, produced this 172 3/8 B&C point "power buck." Jeff Banks proudly shows off the massive rack from the buck known locally as the "Horn Donkey." Jeff killed his big buck on opening day of the 2001 season.

Introduction

By Duncan Dobie

Over the past 20 years, I've written dozens of stories about record bucks in Georgia. It's always exciting to meet a deserving hunter who has taken a record buck and listen to his story. It's even more exciting to write it down so that others can read it later on. When I first met Jeff Banks in February 2002, I knew I was talking to a unique individual. Jeff had killed a Boone and Crockett buck on opening day of the 2001 rifle season scoring 172 3/8 typical points, and I was there to write a story for **Georgia Sportsman** magazine about his accomplishment. I had heard about Jeff Banks from Daryl Kirby, an editor at **Georgia Outdoor News**. Daryl and Jeff are long-time friends. For several years prior to the 2001 season, Daryl had told me some things about the Banks Farm and the incredible trophy management program that was taking place there. I knew that I would eventually cross paths with Jeff Banks sooner or later.

In 1999, when I wrote the story about the Buck Ashe trophy being returned to Georgia from its long exile in Oklahoma and being declared a new Georgia state record scoring 191 4/8 typical B&C points, Daryl Kirby told me that the Buck Ashe trophy had once hung in the Banks home. Since I was so closely tied with the Buck Ashe story, this was an amazing coincidence. (See Chapter 2 for more details regarding Jeff's dad Lamar Banks and how that amazing coincidence came about.)

The first thing that struck me about Jeff Banks was the fact that he refused to take all the credit for killing the big Morgan County buck known as the Horn Donkey. A former college football player, Jeff insisted it had been a team effort all the way with all of the guys who hunt with him on the Banks Farm. Jeff also told me something else.

"I want to write a book about what we've done here on the Banks Farm," he said. "We've learned so much about food plots and protein. I want to share some of what we've learned with other hunters around the Southeast. I think we can help other hunters do the same thing we've done."

Then he paused.

"There's one problem, though," he added. "I'm a deer hunter, not a writer."

The more we talked, the more I was taken in under Jeff's spell. Truly, he and the guys he hunts with have done some amazing things on the Banks Farm. To sum it all up in one simple sentence, they've taken the concept of quality deer management to a new level. Jeff asked me to help him write this book, and I gladly accepted. He went out and bought himself a good tape recorder. We sat down together during the spring 2002 turkey season for several long interviews and recording sessions. **Producing Power Bucks** is the result of those long hours of talking.

Jeff likes to describe himself as a simple deer hunter. Since I'm a simple writer, we seemed to reach a happy medium right away. Even though I wrote all of the words

Introduction

in this book, they are mostly Jeff's words. I wrote Chapter 1 as a lead-in to the book, but everything else is Jeff's original material. I learned a lot while we were working on this book. If you read it, you'll learn something, too.

I'm very proud to be a part of this book because I think Jeff has a world of good information to share with other hunters. In fact, when I think about Jeff's philosophy on supplemental feeding and high-protein food plots, I am reminded of a quote I saw in another excellent book about deer management that was written back in the '70s.

> "There is much argument on food supply to size of deer. I do not think so more that a fat deer is larger than a poor one, but it does govern size of horns where there is abundant feed all year you see larger horned bucks, and hard years smaller horns. A deer taken and raised as a pet and fed plenty of grain will often have 10 points the second year."
>
> Alfred Gardner, State Game Warden
> Laredo, Texas 1939

I think it's fitting that as far back as 1939 a Texas game warden would recognize the important relationship between food supply and the size of a buck's antlers. The quote comes from the book ***Producing Quality Whitetails*** by Al Brothers and Murphy E. Ray, Jr. First published in 1975, this book is today considered to be one of the all-time classics on whitetail management.

The concept of quality deer management got its roots in Texas, and the entire continent has benefited from the pioneering efforts made by giants in the whitetail industry like Al Brothers and Murphy Ray. During the more than a quarter of a century that has passed since ***Producing Quality Whitetails*** was first published, whitetail management has made great strides in every area. By far, nutrition is one of the most important areas of all. In the past, it's also been one of the most neglected.

While working on this book in the spring of 2002, I grabbed several old "how to" deer hunting books off my book shelf that were published in the late '70s and early '80s. Although they were full of good information about scouting, setting up stands, the rut, and other important topics, not one book contained a single word about deer nutrition. Good nutrition and trophy bucks go together like peas in a pod, but only recently have we really zeroed-in on this subject.

When you first read the title to Jeff's book and allow it to sink in for a few moments, two questions will probably come to mind. What is a "power" buck? And what is the Banks Farm way?

The first question is easy to answer. Simply look at the picture on the front cover, and I think you'll agree that any whitetail buck in North America that tallies up a gross score of 197 inches of bone on his head will definitely qualify as a "power" buck. If you need further convincing, take a look at some of the photos of big bucks inside the book. All of these exceptional bucks were taken on the Banks Farm during the past 10 years.

To find the answer to the second question, start reading. You're in for a real treat. It doesn't matter whether you own or lease the land you hunt on, and it doesn't matter whether you control a large tract or a small tract. This book is full of ideas and tips that can help anyone take positive steps to improve their deer herd.

This book is not going to be a lesson in whitetail biology. Jeff Banks is not a biologist. By his own description, he's just a simple deer hunter who has a passion for wanting to grow bigger and better bucks. The only true science in this book is the science of hard work. It's what Jeff, his family, and his companions refer to as "sweat equity" on the Banks Farm. Through trial and error, and through nearly 12 years of working toward a common goal, Jeff and his guys have produced some amazing results.

What is the goal I'm referring to? The Banks Farm goal was to try to restore average body and antler sizes of bucks taken on the Banks Farm to what they used to be back in the "good old days" of the '70s – before the deer herd in central Georgia peaked out and became too populated for its own good. How would this lofty and difficult goal be achieved? It would be achieved through a cutting-edge trophy deer management program that put special emphasis on high-protein food plots and supplemental feeding.

During the past dozen years, the Banks Farm has evolved into a veritable protein factory for the resident deer herd. Jeff has become both a fanatic, and an expert, on high-protein plantings for deer. In fact, he's so obsessed with the subject that this book was a natural by-product of the past 12 years' experience. Jeff and his club members regularly plant a variety of clovers, oats, alfalfa and peas that provide year-round nutrition for the local deer and turkey population on his farm. In conjunction with their planting program, the Banks Farm boys also maintain a supplemental feeding and mineral program that offers additional protein for the local deer on a year-round basis.

Another thing that impressed me about Jeff Banks shortly after meeting him was his passion for the land he lives on and his passion for chasing and growing trophy whitetails. I've seen this same sort of passion in other deer hunters. It's the real energy force that separates the average hunters from those who achieve great things, and it's contagious! You'll catch it if you're around it long enough. Jeff is not merely a whitetail fanatic. He passed that point long ago. He's obsessed with what he does and what he's been doing for the past dozen or so years. More importantly, though, he has a great desire to share some of his acquired knowledge with other hunters.

To be successful in any endeavor in life, I strongly believe that a man must have a true passion for what he does. This passion has to come from the heart. Coupled with a strong work ethic and the right kind of attitude, it can only lead to eventual success. Through a work ethic taught to him at a young age by his family and continued on during his college football career, Jeff set some important goals for his farm in 1990. Now, he and the guys have begun to attain some of those goals. In so doing, they've taken the concept of trophy management and planting high-protein food plots and supplemental feeding to a new level. But they're not content to rest on their laurels and stop here. They think they can grow some bucks bigger than the Horn Donkey in the future.

In some ways, the story inside this book is a fairy tale story with a fairy tale ending. Although Jeff makes it look ridiculously easy, he'll be the first to tell you that it took a lot of hard work by a lot of different people. As you'll see in some of the chapters in this book, it was a team effort all the way. Lamar Banks, Jeff's dad, sums it up nicely. "In the early days we knew all about farming – how to plant, how to prepare the soil, how to fertilize and so on – what we didn't know was how to plant for deer."

But they learned. Boy did they learn! They also proved an important point. No matter what kind of genetics the deer on your property have, if you can control your deer population through keeping doe numbers down, and if you have plenty of high-protein foods available to those deer on a year-round basis, you can increase the antler size of your bucks by a considerable percentage.

Truly, Jeff and all his companions who hunt with him on the Banks Farm have a remarkable story to tell. If they can do it, so can you!

Duncan Dobie
Marietta, Georgia
May 2002

This daytime photo of the Horn Donkey in the edge of a food plot, taken in mid-September 2001 by a trail camera belonging to Charles Hendrix on Innisfail Farm (adjacent to the Banks Farm), clearly shows the heavy mass on the big buck's 6x6 main-framed rack. Everyone thought the Horn Donkey was an exceptional trophy buck, but no one dreamed he would qualify for the all-time Boone and Crockett Record Book.

Chapter 1

A Legend Is Born

They called him the "Horn Donkey." On film, at least, the over-sized whitetail appeared to have the body of a small donkey and the "horns" of an exceptional trophy buck. Charles Hendrix, a good friend of Jeff Banks, is credited with naming the big buck after seeing photos of him taken by a trail camera. Charles' family owns Innisfail Farm, which shares a common boundary line with the Banks Farm. Neither Charles nor any of the regular "guys" who hunted with Jeff on the Banks Farm could say for certain that they had actually seen this trophy buck in the flesh before. Therefore, no one knew just how big this Morgan County bruiser actually was.

In fact, after studying two or three different photos of the buck's long-tined rack taken by several different trail cameras in several different locations during the late summer and early fall of 2001, Jeff and several of his closest hunting companions underestimated the buck's score by a considerable margin. Perhaps they tended to be a little on the conservative side because they didn't want to be guilty of exaggerating.

In any event, everyone knows that photographs can be deceiving. Most hunters tend to over-estimate the size of a buck's rack in a photo, just as they do if they see the buck standing out in a field somewhere. In this case, however, the majority of Banks Farm hunters were low in their estimates. A couple of hunters thought the big buck might score around 150. Jeff guessed that he might gross in the high 160s. Only one man, Jeff's first cousin Danny Keever, was close. Danny speculated the Horn Donkey might gross as high as 180 points. Then Danny second-guessed himself by saying, "Man, he can't be that big, can he?"

In truth, no one dreamed the Horn Donkey would actually wind up with a gross score of 197 points. But it wouldn't be long before they would all find out. Opening day of the 2001 rifle season would end all speculation once and for all. Jeff and all of the men who hunted with him on the Banks Farm were experienced hunters who were familiar with judging trophy bucks on the hoof, but how could any of them know they were about to hit the jackpot? How could they know that the fruits of their labor for the past 11 or 12 years were about to pay off with the ultimate prize?

On the afternoon of October 27, 2001, deer hunting history was made in Morgan County by a bullet fired from Jeff's rifle. Almost immediately, the Horn Donkey became a local sensation. But it wasn't simply the giant 16-point rack which grossed 197 B&C points that made this buck so unique. Nor was it the six-inch-plus bases on the buck's massive rack that contributed to a net typical score of 172 3/8 B&C points after deductions.

In simple terms, the thing that set this buck apart from most other record-book whitetails taken in the state of Georgia was the "sweat equity" contributed by numerous people that went into the endeavor. Sweat equity is a term Jeff Banks uses a lot. You'll see it over and over again in the following chapters. Although Jeff happened to be the "lucky" hunter who pulled the trigger, he'll be the first to tell you that the Horn Donkey could just as easily have walked by any number of hunters who were hunting on the Banks Farm that day. Everyone knew the Horn Donkey was living in the area, and everyone had an equal chance to bring down this amazing animal.

By some strange twist of fate, though, Jeff happened to be holding the winning lottery ticket that day. However, the long journey that culminated with putting a buck of this caliber in the Boone & Crockett Record Book was no accident. Nor was it a lucky coincidence or a single-handed endeavor by one man. Jeff refuses to take all the credit. He insists that it was a team effort all the way. In fact, a popular expression comes to mind that is entirely appropriate to note here. *It's not the destination that's important. It's the journey.*

Two generations of avid deer hunters — Lamar Banks (left) and Jeff Banks (right) — pose with the by-product of 12 years of intensive trophy management on the Banks Farm. "The Horn Donkey is a direct result of considerable 'sweat equity' invested by numerous people," Jeff says. Without question, the record buck is also a direct by-product of an intensive high-protein planting and supplemental feeding program.

Since joining the Banks Farm hunting club in 1993, Danny Keever, Jeff's first cousin, has only taken one buck off the farm, this high-racked 10-pointer shot in 1998. Although Danny's buck grossed 133 B&C points, making it a legal trophy under Banks Farm standards, the miniature whitetail field-dressed at a meager 101 pounds. Because of its small body size, the deer appeared to be much larger than it actually was. Dubbed the "Pansy Buck" by the other Banks Farm hunters, Danny received much ribbing from his fellow hunters because he had passed up numerous bucks in the 135 to 140 B&C point range in previous years. Because this buck might have only been 2 1/2 years old, Jeff Banks calls it a "genetically superior freak."

By most standards, the destination in this case would be a buck large enough to qualify for the Boone and Crockett Record Book. In Jeff's case, the taking of a buck the size of the Horn Donkey was just an interim stopping-off point. When it happened in the fall of 2001, the journey was far from over.

Certainly Jeff is as proud of his achievement as any hunter could ever be, but the real story here is the journey and the lessons learned along the way. It was the journey that produced the Horn Donkey as well as a number of other "power bucks" taken on the Banks Farm in recent years. For Jeff and all the guys who hunt with him on the Banks Farm, the journey was often long and frustrating. It involved much hard work, but the work was always a labor of love. Great sacrifices were made along the way.

For instance, since joining the club in 1993, Jeff's cousin Danny Keever has shot only one buck. Danny's buck grossed 133 points. Strangely enough, it was a small-bodied animal that field-dressed at a meager 101 pounds. (Today, most mature bucks taken on the Banks Farm field-dress close to or over the 200-pound mark.) Because Danny's buck had such a small body, its rack appeared much larger than it actually was as the buck was walking through the woods. This is a common mistake that hunters make in Texas all the time. Had Danny known the buck had such a small body, he probably never would have pulled the trigger.

Danny has passed up literally dozens of trophy bucks on the Banks Farm since 1993. Many of the bucks he passed up score between 130 and 140 points. Several score over 140 points. Instead of shooting these bucks with a rifle, Danny shoots them with a video camera. He has the video footage to prove it. Why has he passed up so many animals that practically any hunter in the south would be thrilled to shoot? He has an appropriate answer.

"When the right buck comes along, he'll make the Horn Donkey look like yesterday's news," Danny says with a smile.

In all seriousness, Danny knows the right buck *will* come along eventually. And when it does, it won't be a 130 or 140-class deer. "It'll be a sho-nuff power buck!" Danny says. This is the kind of attitude that all of the guys who hunt with Jeff seem to possess.

"Sometimes we have to stand on our rifles when we're out in the woods," most of the Banks Farm hunters will tell you. "But it's well worth the sacrifice. The Horn Donkey proved that."

Yes, Jeff and the guys used a lot of restraint in passing up bucks during the past 10 years, and they also invested a lot of hard work into the operation. They learned how to plant the best high-protein foods through trial and error because they knew they would be rewarded in the end.

In October 2001, the goal they had set out to achieve was exceeded beyond all expectations. When the Horn Donkey finally fell to a hunter's bullet, it was not just a victory for Jeff Banks. It was a victory for every member of the group who had invested years of time and effort into such a unique and innovative management program.

The Horn Donkey was 5 1/2 years old when he died. Jeff likes to remind people that there is no way to estimate how many hunters on the Banks Farm passed him up when he was 2 1/2, 3 1/2 or ever 4 1/2 years old. No one could say for sure that they had even seen him in the past, but the odds favor the likelihood that he was probably seen and passed up on more than one occasion.

That, along with a state-of-the-art nutritional program pioneered by Jeff and the guys, illustrates the kind of dedication it takes to grow true power bucks. For Jeff and the boys, the decade of the '90s was a long and interesting journey. The decade culminated with dozens of fond memories and dozens of stories to tell. Even though the journey continued on and achieved an incredible milestone in 2001, the trip is still far from over as already mentioned. Jeff and the guys still have new frontiers to conquer. And when you're on the cutting edge, anything is possible!

"How about a new state record?" Jeff says with a smile.

All kidding aside, some of the simple lessons learned about supplemental feeding and planting high-end food plots by Jeff and his gang through trial and error can greatly benefit any individual hunter or group of hunters in the Southeast. Read on. You won't be disappointed. And who knows? In a few short years, *you* could produce a true power buck like the Horn Donkey of your own.

My dad Lamar Banks has been an avid deer hunter all his life. Here he poses with a very old buck taken off the farm in 1995. The buck was a main-framed 6x6 but he had very short tines in the front. He was starting to go downhill.

Chapter 2

How We Got Started on the Banks Farm

It all began in the early '70s when my dad, Lamar Banks, and several of his friends purchased a farm just south of Madison, Georgia. A few years later, Dad and my grandmother Catherine Banks bought out his partners and added several smaller tracts to the original farm. Today the Banks Farm covers about 3,000 acres, 800 of which we lease from another landowner.

Dad's been an avid outdoorsman all his life and he's always had an interest in farming. When I was growing up, he was always out hunting or fishing somewhere. In his early days, he frequently hunted deer down at Piedmont National Wildlife Refuge in Jones County and also on some private land off Highway 11 in southern Jasper County.

We lived in Tucker, Georgia, until I was five or six years old. For a while, dad worked at the General Motors Assembly Plant in Doraville, Georgia. But he always wanted to do better for his family. Dad is a self-made man, and he started buying and selling farm land. The first farm he bought was in Newton County.

One day Dad decided he'd finally had enough of city life. He bought a farm in Jasper County and our family moved to the farm. For a boy like me, it was like dying and going to heaven. Living in the country gave me the opportunity to spend a lot of time outdoors hunting and fishing. I killed my first deer, a doe, in Jasper County when I was seven years old.

Coincidentally, while we still lived in Tucker, my dad tells me that I was on the same T-ball team with Mark Ashe. Mark, who's about my age, is the son of Buck Ashe. In 1961, Buck killed the biggest whitetail ever taken in Georgia. He was hunting down in the eastern edge of Monroe County near the present-day site of Rum Creek Wildlife Management Area at the time. Buck was an avid bowhunter, and up until that time, he had never deer hunted with a firearm. On the day he killed his big deer, though, Buck was hunting with a borrowed rifle during the November gun season.

Some friends wanted him to go hunting with them that day, and he didn't have his bow with him. So they loaned him a Marlin .30/30, and he killed a giant buck that later scored 191 4/8 typical B&C points. (Note: Buck Ashes's trophy deer is not only the Georgia State Record typical whitetail, it also ranks as the largest typical whitetail ever taken by a hunter in the entire Southeast.)

Buck's deer was not officially scored and recognized as a state record until 1999. But even back in the late '60s, everyone in the Tucker and Chamblee area knew it was a monster. Buck always regretted shooting his deer with a rifle instead of a bow, and to this day, he still talks about it.

My dad heard about Buck's deer one day in the late '60s. The mother of one of Dad's friends did some babysitting for the Ashe children. Buck and his wife were divorced at the time, but Mrs. Ashe and the children lived in Chamblee. The man told Dad there was some kind of a deer mount hanging in the Ashe home that was

The Buck Ashe deer hung in our home for nearly a year in the late '60s when I was around five years old. Dad was in awe of it. He always said it was the biggest whitetail ever taken in Georgia. Back then, he had no way of knowing that he was right. Thirty years later, the Buck Ashe deer was officially ranked as the largest typical whitetail ever taken in the state.

In 1999, Duncan Dobie located the Buck Ashe trophy in Tulsa, Oklahoma, after a nine-year search. It was brought home to Georgia and officially scored at 191 4/8 typical B&C points. Buck's deer grossed around 210 B&C points. Unfortunately, it had over 12 inches in abnormal "sticker" points and other deductions which brought the net score down to 191 4/8 points. Here, Buck Ashe poses with the newly remounted trophy in 1999.

When I was 10 years old, I was sitting in the dentist's chair in Monticello, Georgia, on a Saturday morning in 1974 when my dad came in and got me. Mr. Tom Cooper had just killed his incredible non-typical "Silver Bullet" buck at B.F. Grant WMA in Putnam County. Dad had bumped into Mr. Cooper at the town square in Monticello, and he asked him to drive over to the dentist's office so I could see the trophy. Mr. Cooper was nice enough to say yes. I'll never forget seeing that giant buck lying in the back of a truck.

so big it had to be either an elk or a moose. Yet Mrs. Ashe claimed it was a whitetail taken by her ex-husband. The rack was so huge that a lot of people didn't believe it could be a Georgia whitetail.

Dad's curiosity was aroused. Being an avid deer hunter himself, and also having hunted quite a bit in the same area where Buck's deer was killed, he had to go see the deer head that everybody was talking about. Dad had killed a lot of deer in his own right, but never anything of trophy size as of that time.

Dad found out where the Ashe's lived and he drove over there and knocked on the door. Mrs. Ashe answered the door. Since Dad and the son of Mrs. Ashe's baby-sitter were good friends, Mrs. Ashe invited Dad inside to look at the deer. That was over 30 years ago, and Dad still gets excited to this day when he talks about it. When he saw Buck's deer hanging on the wall for the first time, he said chills went up and down his back. He was absolutely awe-struck. The big whitetail truly *was* the size of an elk!

Dad was so impressed that he told all of his friends about the rack. Several of his hunting buddies wanted to see it. Dad didn't want to trouble Mrs. Ashe by taking people over to her house, so he found out that Buck had moved to New Orleans and he called Buck on the phone and asked him if he could borrow the deer head for a while to show some of his friends. Dad has always had a special way with people, and apparently Buck thought he was a trustworthy person. The two men had a lot in common, although Buck was about 10 years older than Dad. Buck told Dad he could borrow the deer head for a while. "When I want it back, I'll give you a call," Buck said.

Dad borrowed the deer head and hung it in our house. Amazingly, it stayed there for almost a year. I don't think a day ever went by that Dad wasn't in total awe of it. "Some day, I'm going to kill one like that," he would often say. I was only four or five years old, and I don't really remember much about Buck Ashe's deer hanging in our house. But I do know that some of Dad's passion for big whitetails started rubbing off on me about that time. I didn't know what all the fuss was about, but I knew Buck's deer was bigger than a huge mounted Canada goose Dad had sitting next to the deer head, and that's what impressed me the most!

Dad took some pictures of the mount, and Buck finally called him one day and asked him to return it to Mrs. Ashe. Dad promptly took it back. We later learned that the next time Buck came to town he took the deer head back to New Orleans with him to show some of his friends down there. It still had never been officially scored, and it would not be returned to Georgia until 1999. Dad always knew that Buck's deer had the largest set of antlers he had ever seen in his life, but another 30 years would go by before it would be officially measured and recognized as a Georgia state record.

The next really big deer I remember seeing and being totally impressed with was the Silver Bullet buck taken by Mr. Tom Cooper at B.F. Grant WMA in 1974. I was about 10 at the time, and I just happened to be visiting my dentist, Dr. Tom Brady, in Monticello on that Saturday in November. Somehow, Dad and several of his friends heard about the deer being killed that morning. They bumped into Mr. Cooper while he was in Monticello showing it off. Dad always had a knack for being in the right place at the right time for things like that. The deer was so big and so impressive that Dad asked Mr. Cooper if he'd mind driving over to the dentist's office just west of town so I could see the buck. Mr. Cooper was nice enough to bring the deer by, and Dad came in and got me out of the dentist's chair. Dr. Tom Brady was also a deer hunter, and we all went outside to look at that huge buck lying in the back of the truck. That was a sight I wouldn't soon forget.

Tom Cooper's impressive drop-tined buck scored 215 7/8 non-typical B&C points. It not only won the statewide Big Deer Contest in 1974, but it also won the National Rifle Association's distinguished Silver Bullet Award. Mr. Cooper's trophy buck is the only Georgia deer to ever win that award. Today it ranks as the fifth largest non-typical buck ever taken in Georgia.

We heard about a lot of trophy bucks being taken in Jasper, Newton, and Morgan counties during the early 1970s. The first record-size buck that I ever remember seeing in the wild was in Jasper County in 1973. It was the first week of the '73 season and my mother and I were driving down the dirt road on our farm. It was drizzling rain, and we saw a monster buck standing out in a pasture. Knowing what I know today about mass and tine length, I'm certain that buck would have grossed over 170 B&C points. I had a chance to take him but it just didn't work out.

The late '70s more or less marked the end of the heyday of big deer in Central Georgia. The doe population started exploding about that time, and the body and antler sizes of bucks started going down. I didn't know it at the time, but the deer herd in that area was beginning to peak out and get over-populated. The deer numbers were simply higher than the land could support.

We weren't helping the situation much, either. My family members and everyone else who hunted with us on our farm in Morgan County always wanted to kill a good buck, but we still shot every little buck we saw. We never shot enough does, and we didn't give much thought to nutrition.

The first time I ever remember thinking about trying to improve our deer herd was in 1986. My younger brother Lane and I got into a conversation about the benefits of passing up younger bucks. I think I was home from college during hunting season. I attended Georgia Southern College in Statesboro on a football scholarship where I played outside linebacker and defensive end. One of the highlights of my college career was playing on a National Championship team under Coach Erk Russell in 1986. I think my college football career really helped me develop a strong work ethic and a desire to set tough goals and then try to achieve them.

At any rate, my brother and I had gone out as usual and killed several little bucks and we were mad at ourselves for what we had done. We were tired of seeing small bucks all the time, and we started talking about letting little bucks go. A day or two later, I passed up two different four-pointers in the same morning. Boy, I really thought I had done something. I guess I had because you have to start somewhere. Passing up two bucks in one morning was a milestone for me.

During the mid-'80s, we had a lot of friends and guests who hunted on the farm. They killed quite a few small bucks as well. No wonder we never saw any older bucks.

My first association with green fields or food plots had to do with seeing more deer. If we had a planted field somewhere, we'd always see a lot of deer either in that field or nearby. That was very appealing to me. In those days we leased out our open land to local farmers, and they planted some of the fields in wheat, soybeans and corn. Those fields were always loaded with deer.

Because you'd always see more deer in the green fields, I liked the idea of having some planted fields on the farm that would draw the deer in, especially for late winter grazing. We started planting a little wheat on our own in the mid-'80s, but we still didn't think much about high-protein plantings or what deer actually needed to help them make it through the leaner times of the year.

Around that same time, I started putting out some minerals on the farm. I just happened to run across some packaged minerals at a store down in Statesboro while I was at school. I didn't know much about deer nutrition, but I did know that deer needed calcium and phosphorous for their antlers. The label said that these minerals contained a high percentage of calcium and phosphorous, so I just thought it would be a good idea to put them out. I'm proud to say I've been putting minerals out ever since!

This was my first "trophy" buck – taken on the farm in 1984 when I was 19. He was 4 1/2 years old, and he grossed 140 B&C points. He had been eating soybeans. Bucks like this were few and far between in those days.

Setting Some Serious Management Goals
Five Important Keys

After I graduated from college in June of 1990 and came home for good, Dad and I decided that we wanted to do a better job with managing our farm in Morgan County. I had been hunting on the place since I was seven years old, and it had come to mean a lot to me. The first thing I did was fire up an old front-end loader we had. I did a lot of work on the road system throughout the farm. I put in some fire breaks, and we did some controlled burning in some areas that really needed it. In general, I did a lot of piddling around with that loader. I had a real burning desire to make things better.

In addition to burning, we started cleaning up and reclaiming some old fields that had grown up. These grown-up fields eventually turned into some of our best food plots. I had several friends in the timber business who let me borrow some dozers and other equipment for site preparation and planting. I'd go out into an old field and pick up all the big sticks and limbs by hand. Then I'd get the field ready to plow and plow it up. Finally, I'd get the field in a position so that we could start planting.

By the late '80s, there was a lot of interest in quality deer management. A number of articles had been written on the subject and hunters were beginning to really talk it up. I read everything I could get my hands on pertaining to deer management. We started encouraging everyone who hunted on our farm to pass up younger bucks. We finally implemented a rule that said if you shot it, you had to get it mounted. But our guests were still shooting small bucks.

1990 was the real turning point on our farm. It more or less marked the beginning of our modern deer management program. After learning as much as I could about some the latest deer management techniques, I came up with five basic keys that I thought would be critical to the future success of any management program on the Banks Farm. After listing these keys, I quickly realized that the first two keys had already become serious problems in our part of the state. Those five keys were:

> **1) Shooting does** – Our doe population was exploding and we weren't shooting nearly enough does on the farm. (As already mentioned, during the late '80s, the entire deer population in Central Georgia had increased to the point that the deer herd had exceeded the carrying capacity of the land. Whenever this hap-

After we started planting food plots containing high-protein iron and clay peas during the early '90s, we discovered a direct correlation between the food plots and all the good "power" bucks we were beginning to see.

pens, deer get smaller, and the body and antler size of bucks decrease dramatically.

2) Letting small bucks go — Everybody wanted to kill a big buck on the Banks Farm, but in truth, most hunters usually ended up shooting the first small buck that came by.

3) Planting food plots — We knew we already had an over-population problem. Even though I didn't know much about deer nutrition, I strongly believed that we could make a difference on our farm by planting more food plots.

4) Supplemental feeding — My dad had put out several deer feeders years ago and fed corn at certain times of the year. We later learned that corn is very low in protein. Although we knew next to nothing about high-protein feeds in 1990, we still felt like supplemental feeding was a great idea. We didn't really perfect our supplemental feeding program until the late '90s, but we did start building more feeders and we realized that there were certain times of the year when the deer really needed good supplements.

5) Minerals – Again, I knew next to nothing about the proper minerals to put out for deer. But I did know enough to realize that the right kind of mineral supplements could help a doe raise a healthier fawn. I also believed that minerals could help grow larger antlers on our bucks.

An Unknown Quantity

We started shooting does with a vengeance in the late '80s and early '90s. Maybe that sounds a little strong, but it was something we knew we had to do. In those early days, it was a very difficult job to accomplish because we only had so many doe tags to work with and it was hard trying to shoot enough does without breaking the law. But we were determined to get the job done, and we found ways to do it. We made a few mistakes and we shot a few button bucks. In those early years, around 10 percent of the antlerless deer we killed each year were buttons. We got better as time went along. We regarded shooting a few button bucks as a necessary evil that went with the territory. We knew we had to make some sacrifices in order to get our deer numbers back down to where they had to be.

We also began making a real effort to let the 1 1/2 and 2 1/2 year old bucks go. Those two keys – shooting more does and passing up younger bucks – are the backbone to any serious deer management program. Those were givens. We knew we had to accomplish those two objectives. We also knew we could accomplish those objectives with some hard work. It was just a matter of putting in the time and effort to get the job done. Fortunately, I had a lot of good help. The guys who were hunting with me on the Banks Farm were just as passionate about our deer program as I was. We all knew what the stakes were, and we knew what we could accomplish. There was never any doubt.

Even though a lot of work went into shooting does and a lot of restraint was required for passing up younger bucks, we knew that we'd eventually end up with a healthier deer herd and a healthier range for them to live on. And since we were allowing young bucks to live, we knew we'd eventually start seeing some 3 1/2 and 4 1/2 year old animals.

The real unknown quantity for us was nutrition. We knew that planting food plots would make a difference, but how much of a difference no one could say for certain. I just had this gnawing feeling inside that it would make a huge difference. The more I learned, the more excited I became. I felt that with the right kind of food plots and year-round supplemental feeding, we could actually see a 15- to 20 percent increase in antler size from trophy bucks taken off the Banks Farm. In truth, what I really wanted to accomplish was to try to get our deer herd back to what it had been in the 1970s. Back then, mature Morgan County bucks frequently field-dressed at over 200 pounds, and their antlers could score anywhere from 130 to 150 B&C points on average.

In other words, if one of our 3 1/2 year old bucks in 1990 was capable of growing a rack that grossed 130 inches, a 20 percent increase would put him over the 150 point mark. That's what I wanted to do and that's what I honestly believed we could do in a few years time. But I had nothing to back it up. It was only a gut feeling. We

1994 turned out to be a banner year for big bucks on the Banks Farm. I killed my biggest buck ever, a beautiful 12-pointer that grossed 161 2/8 B&C points. Because of all the rain that year, the deer on the farm had more iron and clay peas than they could possibly eat.

were just getting started, and I knew it would take a minimum of several years to start seeing any positive results.

Dad had been planting small food plots on the farm as far back as the early '80s – mostly in wheat. But 1990 was the year we really got serious about nutrition and high-protein plantings. We realized that wheat alone didn't offer our deer the kind of high-protein feed that other planting like clover, peas and alfalfa offered. We also realized something else. In order to stay in top condition, deer need high-protein foods 365 days a year.

If deer get high-protein plantings in the spring and summer, that's great, but what about those lean months in January, February and early March? What about late summer when much of the browse is gone and the acorns are not yet falling? Food plots alone can't provide high-protein foods 365 days a year. With the right kind of supplemental feeding and minerals, though, filling in those gaps is possible.

This is what I believed, and this is what I set out to try to accomplish. Furthermore, if the deer are not overpopulated and the woods they live in are in good condition, there are many natural foods like browse and acorns that they'll eat as those foods become available. Deer don't eat just one type of food all the time. During a 24-hour period, they travel from place to place and they eat many different types of food along the way. I wanted every buck, doe and fawn on the Banks Farm to be in top condition, and the only way this could happen was to offer them the best high-protein plantings and supplements seven days a week, 52 weeks a year.

Since most of our late winter food plots had always been planted in wheat, I wanted to try something different. I had always heard a lot of good things about iron and clay peas and how much deer were attracted to

Taken on November 12 during the peak of the rut, my '94 "power" buck later appeared on the December 1994 cover of Georgia Outdoor News.

them. I knew Pennington Seed, Inc. in Madison sold them, so in 1990 I purchased several hundred pounds of seed and planted them in several of our summer food plots. At the time, I had no idea what their protein content was. I just knew that the deer really liked them and they were supposed to help fatten deer up. Several years later when I started learning more about the importance of protein, I sent some of those iron and clay peas off to be tested. I found out they contained anywhere from 21 or 22 percent all the way up to 27 percent protein. That's pretty doggone good for a deer planting.

We couldn't plant our iron and clay peas much before May. In the mean time, between January and the end of May our deer weren't getting much protein from the winter wheat we had planted. This is part of that gap in the calendar I was talking about earlier. I didn't know it at the time, though. I was just making sure they had something to eat. It took several years into our management program before we realized that we needed to have good, high-protein foods available from January through December – either through plantings or supplements.

An Important Discovery

After my first experience with iron and clay peas in 1990, I continued buying them from Pennington Seed, Inc. and planting them every spring. I knew deer loved soybeans as much as they loved peas, and soybeans are high in protein. But I quickly learned that when a deer picks the leaf off a soybean plant, new leaves don't grow back. With pea plants, however, as long as the plant is getting enough water it'll keep putting out new leaves. That just puts more tonnage out there for the deer for a longer period of time – as long you're getting sufficient rain.

In all, four big bucks were taken off the farm in 1994. Here, Bruce Coleson shows off his wide-spreading main-framed 4x5 trophy. Despite a missing brow tine, Bruce's buck still grossed around 149 B&C points.

We had a very dry year in '93. Early in the summer, we saw some nice bucks in velvet out in our food plots containing iron and clay peas. In late summer, though, after the peas started burning up due to the drought, we stopped seeing deer in those food plots.

The following year, 1994, something very interesting occurred. I was out planting peas with some of the guys on either the first or second day of May. When I refer to the "guys," I'm talking about the guys I hunt with on the farm. During the early '90s, we formed a very close-knit group of about 10 or 11 hunters. Some of the guys are cousins of mine and others are college friends or just close friends in general. But our group is very compatible, and it's also very family-oriented. Everyone in the group contributes to the overall program in some important way. I'll go into more detail about that in a later chapter.

Here's another 1994 power buck. While hunting with his father, Royce Grant shot this beautiful, "picture-perfect" 10-pointer that scored 145 points. You could tell this buck had been eating some high-powered groceries!

At any rate, we were all out breaking our necks to get the peas planted because heavy rains had been predicted, and we knew they were on the way. We were desperately trying to get our seed in the ground before it started raining. We had our fields fairly well prepared, and thank goodness we finished planting just as the rain started. I'll never forget it. Just as I started driving home in my truck that day the heavens opened up. It started coming down in buckets.

Once the rain started, it didn't stop until August. It was one of the wettest seasons we could remember. In fact, it rained so much that a lot of flooding occurred around the state. But our peas loved the rain. We had peas in some of our fields that were waist high. You wouldn't believe the deer we saw in those fields that summer once the peas were up. We saw dozens of healthy looking does and fawns and we saw some unbelievable bucks in velvet. We knew there had to be a direct relationship between the peas we had planted, the heavy rains and all those good bucks we were seeing.

1994 — A Banner Year

The proof-in-the-pudding came that fall during hunting season. 1994 turned out to be a banner year for big bucks on the Banks Farm. I killed my biggest buck ever, a beautiful 12-pointer that grossed 161 2/8 B&C points. We actually killed four big bucks on the farm that year including mine.

Because of all the rain, the deer literally had more peas than they could eat. We had been shooting a lot of does each year up to that point, and we were beginning to get our total deer population down where we wanted it. We had also been doing some burning and that created a lot of natural browse.

On the morning of November 12, I was hunting in an area where former club member Ken Brown had seen a pretty good buck several days earlier. I was sitting in a ladder stand when two young bucks came up in front of me and started fighting. They really went at it – knocking each other down on the ground and hitting their antlers together. After that it got real quiet. I decided to go and sit in another permanent stand about 200 yards away. I more or less still-hunted over to the other stand, and when I reached it, I decided to climb up and sit in it for about 30 minutes.

I had only been there about two or three minutes when I looked behind me and saw a doe that appeared to be all hunched up. At first I thought there was something wrong with her, but it didn't take long to realize she was in heat. I saw another deer's leg right behind her, and I thought it might be one of her fawns. All of a sudden I started seeing some long tines sticking up above the saplings where they were standing, and I knew that second deer wasn't a fawn. Instead, he was the biggest buck I had even seen in the woods on the farm and I was excited!

A few seconds later the buck started walking across the ridge toward another doe and he went out of sight. I almost panicked. I didn't know what to do. I considered climbing down and trying to circle around him and cut him off like I sometimes did while turkey hunting. But that first doe was still there, and I decided to just sit tight and try to keep my cool. All at once, the doe he went toward ran down the hill in my direction. A few seconds later, I could see those tall tines of his antlers coming toward me.

He came out in a logging road and continued walking directly to me. I shot him at 44 yards. He had a wide, beautifully symmetrical 5x5 basket rack with one sticker on each G-2 which gave him a total of 12 points. He grossed 161 2/8, and he later appeared on the cover of the December 1994 issue of ***Georgia Outdoor News***, a statewide hunting and fishing magazine. I was really proud.

In addition to my 12-pointer, Ken Brown shot a beautiful buck that grossed 160 5/8 B&C points. Ken was hunting over a food plot, and eight or nine deer were feeding in the field at the time including several small bucks. All of a sudden a big buck came charging out of the woods into the field. He was chasing a doe, and all of the smaller bucks spooked and ran. The other does just sort of stood there looking at the buck. Ken said he had never seen anything like it. The buck stopped. Although Ken was pretty nervous, he managed to get his gun up and get off a perfect shot.

Another one of our former members, Bruce Coleson,

killed a beautiful eight-pointer with a broken brow tine that would have scored in the 150s if he'd not had that broken tine. Even with the broken tine, Bruce's buck still grossed around 149 points. On the day he killed his buck, Bruce was planning to hunt on the lower part of the farm, but I talked him into going over to an area that had been clear cut several years earlier. The clear cut was rapidly growing up and I knew this would probably be the last year we could hunt it effectively.

I had already killed my big buck, and I offered to go with Bruce and do some rattling and videoing while Bruce hunted. We videoed several nice young bucks that were probably 2 1/2 years old, and then I had to leave to go to the bank for a meeting. I left Bruce in his stand, and about an hour later I got a message on my phone that said he killed a big buck out in the clear cut. I went out there with several other club members and helped him drag it out. In addition to the broken brow tine, Bruce's buck had a 1 1/2 inch long G-4. He actually would have been a 4x5 if his brow tine had not been broken. Bruce's buck had long beams and excellent mass. We aged him at 5 1/2. We had found one of his sheds at 3 1/2, but we never saw him again until the day Bruce killed him. When Bruce got his buck mounted, he had that shed antler mounted on the same plaque with the deer head.

The fourth big buck taken in '94 was taken by a young hunter named Royce Grant. Royce was probably in his early teens at the time. He was hunting on the railroad tracks with his father near a large clear cut where he and his father had seen a big buck earlier in the season. (A railroad runs through our property.) They both said the deer was much bigger than my 12-pointer and Royce's father believed the buck might make the record book. They never saw that buck again, but they did see a beautiful 10-pointer that Royce ended up shooting. Royce's buck was 3 1/2 years old and he grossed 145 B&C points. Genetically, he was a perfect 10-pointer. You could tell that he had been eating a lot of high-protein feed.

In addition to the four bucks we killed that year, Danny Keever found a fifth buck lying dead in a creek while he was scouting in early October. We suspected that buck might have died from hemorrhagic disease. Carl Wilson and I had seen that buck in July and Danny had actually filmed him in August. He was a beautiful 10-pointer that grossed 150 7/8. We figured he was either 3 1/2 or 4 1/2 years old and we hated to lose a buck that nice to some unknown cause.

We also saw several other big bucks during the '94 season that never got shot. By the time the season was over, we had witnessed first-hand what those peas could do for our bucks in a year's time, and we were already planning to plant a lot more peas the following year.

In 1993, my good friend Jarrod Brannen of Unadilla, a former college friend and a member of our club, killed a beautiful 10-point buck on the farm that grossed in the mid-150s. Jarrod's buck was really the first big buck to come off the farm since we started our management program in 1990. When Jarrod killed his buck it was a big event for everyone in the club. For us, that's when the rubber really started to meet the road. We were all so proud of ourselves because we had worked very hard to try to grow some decent bucks on the farm and it looked like things were finally starting to pay off. We were also proud for Jarrod because his buck was such a beautiful trophy. We had no idea what was in store for us the following year.

Shooting a 150-class buck anywhere in North America is a feat. But doing it in the Southeast is a huge feat. It encouraged us to set our goals as high as possible. Furthermore, it showed us that if we kept fine-tuning our management program, we definitely could grow some 150- to 160-class bucks.

Because of the rain that year, our food plots were in top condition. It was true that we had been passing up younger bucks since 1990, and maybe it was time to start seeing some older age-class bucks. But I was convinced that our food plots had as much to do with helping those bucks grow bigger antlers as anything else we had done. I was really fired up. If the peas helped us grow 160-class bucks in just five short years, what would the future hold? I was already thinking about the "impossible." Could we actually grow a record-book buck that would score 170 or more B&C points? If we offered our deer the right kind of nutrition, I was convinced we could!

Summary — Chapter 2

- **There are five important keys that should be a part of any serious trophy management program in the Southeast. Those keys are:**
    1) **Shooting enough does**
    2) **Protecting younger bucks**
    3) **Planting high-protein food plots**
    4) **Supplemental Feeding**
    5) **Putting out minerals**

- **On the Banks Farm, we realized there was a direct correlation between high-protein plantings and growing bigger bucks.**

In addition to age and good genetics, it takes three things to grow a true power buck like this heavy-horned 136 B&C point 5x5 – nutrition, nutrition and nutrition. Taken by Lamar Banks in 2000, this Banks Farm heavy-weight tipped the scales at 215 pounds (live-weight).

Chapter 3

Nutrition, Nutrition, Nutrition

If you're a serious trophy hunter, I'm sure you've heard or read many times that three major ingredients are necessary for growing big bucks – age, genetics, and nutrition. By protecting smaller bucks – in our case, 1 1/2 year olds and 2 1/2 year olds – the age problem was relatively simple for us to remedy. We started off protecting our 1 1/2 year old and 2 1/2 year old bucks. As time went by and as our program got more refined, we started passing up 3 1/2 year olds and in some cases 4 1/2 year old bucks as well.

It took a lot of restraint to do this as mentioned, and it still does – especially when you begin passing up bucks that score in the mid-130s or higher. But if you ever expect to shoot a 150-class buck or better, you'd better be able to "stand on your rifle" when it's necessary and let those 130-class bucks walk. Anyone can do this. Attitude has a lot to do with it. We've been very successful and very fortunate in this endeavor because the members of our club work together extremely well. We are constantly encouraging each other to pass up good bucks. When one of our hunters does pass up a good buck and chooses to get video footage instead of killing it, the other guys are always there with praise and support.

Since our deer in Morgan County are known to carry excellent genetics, we knew that two out of the three "big buck" ingredients were already present in our area. (Back when the state was restocking the area with deer in the late '50s and early '60s, Wisconsin deer were released in several Central Georgia counties. These Wisconsin whitetails, the largest bodied and largest antlered whitetails in North America, slowly expanded their range into Morgan County.) We felt like the genetics part of the formula would take care of itself. All we had to do was to work on getting our deer population back down to a reasonable level by shooting does.

Speaking of genetics, I firmly believe that even if the deer in your area don't have the greatest genetics in the world for growing big antlers, you can still make a huge difference through supplemental feeding and by planting high-protein food plots. I'm convinced that you can significantly increase the size of your bucks on almost any range in the Southeastern United States with the right kind of nutritional program. Furthermore, I'm convinced that you can increase the size of your bucks' antlers – both mass and tine length – by anywhere from 15- to 20- percent or better through year-round high-protein feedings.

How Many Does Should You Shoot Each Year?

That's a question I get asked a lot. There's no simple answer to this question because there are so many variables. You certainly don't want to shoot too many does, but in all honesty, that would be a hard thing to do. However, from what I've seen in various parts of Georgia, the vast majority of hunting clubs never shoot enough

We started shooting does with a vengeance during the late '80s and early '90s because we knew we had a serious over-population problem. It was not a pleasant job but we knew it was necessary. Here I'm posed with some of my teammates from Georgia Southern University during the late '80s on doe day. We always found a good home for any surplus venison.

does each year. There's no question that many of the Deep South states including Georgia, Alabama and Mississippi have struggled with over-population for many years now.

To get good advice, talk to your local game warden, DNR biologist or even a private whitetail consultant. These professionals can usually give you a good idea of how many deer you have per square mile, and how many does you should try to shoot each year to keep your total deer numbers where you want them to be. Set a goal and try to attain it. It doesn't do any good to shoot a lot of does one year and lay off the next. In most cases, shooting does is a continuous job that has to be performed every single year.

When we started our management program in the early '90s, we knew we had a serious over-population problem. Our goal was to try to reduce our deer numbers until we started seeing results. We have 3,000 acres, and some years, we shot 150 does or more. That's around five does per every 100 acres. As a rule of thumb, some clubs in Georgia try to shoot a minimum of one doe per every 100 acres. But like I said, there are many variables to take into consideration.

I can't begin to tell you how difficult it was for us to shoot 150 does per year, but it was something we knew we had to do in order to get our deer herd back down to a manageable level. Our club members worked together to find ways to accomplish that goal. As time went by and as we started seeing signs that our total deer numbers were getting more in line with our management objectives, we knew we didn't have to shoot as many does as we had in the past.

Hunters for the Hungry Programs

When you shoot as many does every year as we were forced to shoot during the decade of the '90s, finding a good home for all that extra venison can be a serious problem. We do it in a number of ways, and we make sure we never waste one single pound of venison. In addition to our club members and their families taking a lot of the deer meat, we found friends and neighbors in our community who were delighted to take several does each year. Some of these neighbors do not hunt, and I can't tell you how happy they are to get a supply of delicious "protein-filled" venison each season. We have one neighbor alone who accepts nine or ten does every year.

"Hunters for the Hungry" programs are becoming more and more popular throughout the Southeast these days. In Georgia alone, the HFTH program has helped feed thousands of people in need during the last eight or nine years. The way it works is this: On designated weekends during the season, hunters drop off field-dressed deer to designated locations. The venison is then processed and distributed to local food banks.

Alan Hunt poses with a Morgan County bruiser shot on Innisfail Farm in 1965 by Jimmy Hendrix, father of Charles Hendrix. The buck's antlers grossed in the 150s. Although the big buck is only a main-framed 4x4, he has exceptional mass, typical of mature bucks back in that era. By the end of the '80s, big bucks like this were rare in Central Georgia because of over-population. When we started our trophy management program in 1990, our objective was to grow some bucks like we had seen in the "good old days."

If you have a lot of excess venison, check out this program or one similar to it in your area. Also, more and more food banks are taking venison directly from hunters. Do your homework and find a good home for all of your extra venison. It'll not only help feed people in need, but it'll also go a long way toward helping hunter/non-hunter relations.

The Unknown Quantity Becomes an Obsession

The third major ingredient necessary for growing big bucks, deer nutrition, was the one area that we didn't know much about. But like I said, this was the one area where we felt like we could make the biggest difference in trying to grow bigger and healthier bucks with heavier antlers.

Dad knew a good bit about farming, and that was a definite advantage in getting our food plot program off the ground. Dad knew all about how to prepare the soil, and how to plant and fertilize. He also had a good working knowledge about most of the other things that go along with any farming operation. What none of us knew much about was how to plant for deer. I decided it was time to find out.

For instance in the early '90s, I knew next to nothing about lime and fertilizer. I knew lime was important but I couldn't tell you what it did or how it affected the soil. Since Dad had been around farming most of his life, he taught me a lot. He explained that lime helped neutralize the acidity rate in the soil so that the plant could absorb more of the fertilizer. Obviously, if a plant absorbs the right amount of fertilizer, it'll be a healthier plant and it'll reach its maximum potential.

I learned a lot of things like that from Dad, and I also did a lot of reading up on the subject. Then I sat down and tried to figure out how much lime and fertilizer we needed to put out per acre in different areas and in different types of soil.

We never had any professional help from anyone in those early days, but I got a lot of good advice from Mr. Brooks Pennington who owned Pennington Seed, Inc. in Madison, Georgia. As mentioned, I had been buying all of my iron and clay peas from Pennington Seed. For some reason, "Mr. Brooks" seemed to take a real liking to me. He always helped me out whenever I had any kind of problem. Today, it would be almost impossible to reach the corporate president or CEO of a company like Pennington Seed on the telephone, much less talk to them in person. But Mr. Brooks was always there to answer questions. That's the kind of man he was.

After the incredible year we had in '94 – a year in

which I'd personally killed my best buck ever and three other 140- to 160-class bucks were taken on the farm – I was really fired up. I wanted to do even better than we had done in '94. In fact, I *knew* we could do better, and I wanted to plant a lot more peas.

Before it was time to plant in the spring of '95, I sprayed some Round Up out in several of our food plots to help control the weeds and to help get the fields ready to plant. I was so anxious to get those peas in the ground that I didn't wait long enough after spraying the Round Up. I turned those fields over and planted my peas several days too early. I found out the hard way that Round Up and peas don't mix. I should have waited three or four more days before I planted. I burned up most of my iron and clay peas and I was just sick.

I went back over to Pennington Seed and told Mr. Pennington what had happened to my crop. He gave me one of his booklets about planting and told me to read it.

We knew that good nutrition had to be the key for growing bigger bucks on the Banks Farm. Once we learned a little bit about deer nutrition, our goal was to feed our deer a high-protein diet. We did this in three ways – high-protein food plots, supplemental feeding and by putting out minerals.

Then he did something I'll never forget as long as I live. He said his company would replace all the seed I had lost. Pennington Seed, Inc. has always had a money-back guarantee on any of their products, but in this case it was my fault completely. Mr. Brooks knew that I had burned up all my peas with weed killer, but that was the way he did business. He certainly didn't have to do that, and it really made an impression on me. Sadly, Mr. Pennington died on May 23, 1996 after fighting cancer for several years. I read the booklet he gave me cover-to-cover several times and it was full of good information. It told you what to do, and what not to do.

Since that time, I've bought nearly all of my seed for the farm from Pennington Seed, and I've never had any problems. I believe you get what you pay for and everything we bought from Pennington Seed was always top quality. Pennington Seed sells their products nationwide and I heartily recommend them. During the mid-'90s, in addition to our wheat and iron and clay peas, we started planting some alfalfa and some ladino clover in some of our food plots. This was before I knew anything about the yuchi arrowleaf clover that I'm so high on now. I found out the hard way through trial and error that ladino clover is not very drought tolerant, and it doesn't do well on hillsides and high places that are well-drained. Ladino is a bottomland clover and it needs to be in the moist bottoms where you have a little shade. Yuchi tends to be a lot hardier and it'll grow in tougher areas like hilltops and openings in the woods used for loading docks. It's also very disease resistant. I eventually learned that yuchi produces more tonnage than the ladinos and it contains the highest protein level of any clover that I've planted. I'll get into more detail on that later.

Developing the Five Keys to a Higher Level

The early '90s was an exciting time for our club. The five management keys mentioned in Chapter 2 sort of evolved out of our desire to make a difference on the Banks Farm. We were shooting does, letting young bucks go, and planting more and more food plots. Those were the first three keys. The last two, supplemental feeding and putting out minerals, were also being tested.

We were learning new things all the time by trial and error. By the mid -'90s, we felt like we were really beginning to move up the ladder, particularly in 1994 after having had such a great season. Most of the guys in the club had taken at least one good buck by then, and we were now beginning to let 3 1/2 year old bucks and in some cases even 4 1/2 year old bucks walk. It took us a while to get to that point.

We established a one-buck limit per hunter per year early in our program. I believe that is an essential fact of life for any serious trophy hunting program. Even with the one-buck limit, many of our hunters chose not to shoot a buck during any given season. Some of the guys were even going several years without shooting a buck. Again, this is what it takes to become a serious trophy hunter, and that was the mindset of our club. Unlike many clubs, our guys didn't (and still don't) feel like they *had* to shoot a buck each year just to get their money's worth. Instead, these hunters would much rather pass up a "nice" 3 1/2 year old 10-pointer with eight-inch tines – knowing that next year or the year after, that same 130-class 10-pointer could well be a 150-class "power" buck. We constantly encourage each other to pass up good bucks as mentioned, and it has really paid off.

Traditionally, we do all of our serious buck hunting

Filling in those "protein gaps" during certain times of the year is so important, especially during dry years. One of the best ways to do that is by supplemental feeding from January through August. Putting out minerals is also extremely important.

in October and November – before Thanksgiving. After Thanksgiving, we start concentrating on shooting does. We invite fathers and sons, daughters and wives, and other relatives of club members to hunt with us. The Banks Farm has always placed heavy emphasis on family participation. Family members are always treated just like they are regular club members.

Guests of club members are allowed to shoot does with one exception only. If any club member has not shot his one legal buck before Thanksgiving, he can still shoot a good buck after that time if he sees one in the woods. Or, if he so chooses, he can let one of his relatives like a son or daughter shoot his designated buck. This policy of allowing guests to hunt with us after Thanksgiving has worked very well for us over the years. It helps create a real family hunting atmosphere at the Banks Farm which everybody likes. It's also a great way to help reach our doe-shooting goals each year. Perhaps most important of all, it's a great way to introduce youngsters to deer hunting. Over the years, a lot of children, cousins, and nieces and nephews of club members have taken their very first deer on the Banks Farm.

Supplemental Feeding – Filling in the Gaps

Back around 1980, Dad built several wooden feeders on the farm specifically for deer. Some of those old feeders are still around today. We laugh about it now because those first few feeders were built with wooden dividers in them. Even though Dad's intention was to give the bucks on the farm a little supplemental feed during the winter months when there was very little natural food available in the woods, there was no way a mature buck with a full set of antlers could put his head down and eat from one of those feeders. Of course, if he had already shed his antlers it was no problem.

In those days Dad couldn't get any type of deer feed locally because no one sold it. He did some checking around, though, and he called the Purina Company. Purina actually shipped him some deer feed (or goat feed) from somewhere in Arkansas. Dad loved to tell people that he had the first supplemental feeding program east of the Mississippi River, but I'm sure other people were also feeding deer at that time. He was one of the first people, if not *the* first in our area, however, to feed deer back in those days. It was just sort of a hit or miss thing back then, though, and we really didn't develop a serious supplemental feed program until years later.

In 1995, my wife Michelle and I built a house on the farm and moved to the property. I put a feeder in my front yard just so that the children could watch deer from the house. That was really the beginning of our feeder program. After that, we started building our own feeders

Trophy management works on small tracts, too! This massive 10-pointer, taken in Morgan County, Georgia, during the 2001 season, netted 164 typical B&C points. The giant buck was killed by Shane Casper on a 40-acre tract. Next door to the 40 acres, a 140-acre tract was being intensely managed for big bucks by Billy Young, one of our club members. Obviously, this 164 point power buck benefited from the management efforts next door.

and putting them out in various places throughout the farm.

Even though 1994 had been such a great year for us because of the heavy rainfall that produced a bumper crop of peas, our luck seemed to change in the years immediately following. From '95 on we had several very dry "drought" years with little to no rain. Because our food plots didn't do nearly as well as they had in '94 – by late summer, they were literally burning up – we realized very quickly that supplemental feeding on a year-round basis was a necessary and vital ingredient for our overall program. It didn't take us long to figure out that during dry times we could put out feeders and supplement the deer when other food sources were either gone completely or getting very scarce.

Even back in 1980 when Dad built those first couple of feeders, his idea was to give the deer a little help in late winter when very little browse was available. But now we were learning that gaps can occur at other times of the year as well – even in mid-summer – and our aim was to help fill in those gaps whenever they occurred. Supplemental feeding was the way to do it.

Today, we use a high-protein deer feed supplied to us by Godfrey Feed and Seed in Madison, Georgia. Godfrey Feed and Seed is owned by Candler and Whitey Hunt, two uncles of my good friend Charles Hendrix, whose family owns the 2,500 acre Innisfail Farm next to us. Charles, an avid trophy hunter in his own right, is responsible for giving the Horn Donkey his nickname as mentioned in Chapter 1. Fortunately for us, Charles and his various family members have been great neighbors to the Banks Farm because they have been managing Innisfail Farm for trophy deer for many years just like we have. In fact, the two farms work together in many areas, and our management programs complement each other. Together, we have about 5,500 acres under trophy management.

Recently, Godfrey Feed and Seed started selling a specially blended deer feed that contains a slightly higher percentage of protein than their original feed. We've been using that feed, as well as a very nutritious mineral supplement also sold by Godfrey Feed and Seed, very successfully.

Mineral Supplements

As mentioned, I started experimenting with putting out minerals during the late '80s while I was still going to college. From that time on, I always believed that mineral supplements were an important part of the formula for growing big bucks. I didn't know much about minerals, but I did know that salt alone, with just a few trace minerals added in, didn't do much to help the deer. From the time I was small, I could remember people talking about putting out salt licks, and we had several on our farm years ago.

What I wanted was just the opposite. I wanted something with good mineral content and just enough salt to attract the deer. My theory about minerals was based on common sense. I knew that doctors recommended pre-natal vitamins for pregnant women. In my mind, it stood to reason that if pre-natal vitamins helped make human babies stronger and healthier, it made sense that vitamin and mineral supplements could do the same thing for does when they are raising fawns. The healthier a male fawn is when he hits the ground, the better chance he has of growing up to be a bigger and healthier buck. That goes for the mother as well and her doe fawns. I also believed that, ultimately, minerals could help mature bucks grow better antlers.

We've been putting out mineral supplements on the Banks Farm since 1986 now. Although some biologists say that the deer only receive about 25 percent of the total nutrients they take in, my theory is that we've been doing it for so long that they ought to have plenty of it in them by now. I think you can look at the antler mass on some of the bucks we've taken off the farm in recent years and see that *something* has made a difference. Some of our racks have seven-inch bases, and a lot of our racks have six-inch bases. I believe that's due to the combination of minerals and protein available to our deer 365 days a year.

By making a variety of supplements available to our deer – plantings, feed and minerals – the deer on the Banks Farm don't ever have to go far to find one of these supplements. No buck is going to stay at the mineral lick all day long. And he's certainly not going to stay at the feeder all day long. He won't stay out in a food plot all day long either. But he may cruise around from place to place and eat a little bit here and a little bit there. That's the important thing to remember. To grow real power bucks, deer have to have a combination of each of these ingredients available to them 24 hours a day, 365 days a year.

Large Tracts versus Small Tracts

I realize I've been doing a lot of talking about the Banks Farm and what we've been able to accomplish on 3,000 acres. We are also extremely fortunate because we have several large tracts around us like the Innisfail Farm next door practicing the same type of intensive management we practice. Many people believe that you have to have several thousand acres in one block to make any serious attempt at quality deer management practical. I don't think that's necessarily true. It's nice to have a large tract. We are very fortunate to have a large tract. We are also fortunate to have some large tracts around us where the owners are practicing quality management on their land as well. But I firmly believe that if you practice the five keys mentioned in Chapters 2 and 3, you can make a huge difference on a small tract as well – whether you own or lease.

Billy Young, one of our club members and a very knowledgeable hunter, has been looking after a 140-acre tract for a man in Morgan County for several years now. Their management program is very similar to ours. They've been planting high-protein food plots, they've been taking plenty of does, and they've been letting younger bucks go. During the 2001 season, the same year I killed my Boone and Crockett buck, Shane Casper, who owns a 40-acre tract next to the 140 acres, shot a beautiful 10-pointer that netted 164 B&C points. You can't convince me that the management program on the 140 acres didn't have something to do with growing that 160-class power buck – even on a tract as small as 40 acres.

If local landowners can produce those kinds of results in Morgan County, Georgia, there is no reason why anyone anywhere in the Southeastern United States cannot do the same thing. The potential is there. You simply must have the desire to do it and you've got to stick with the program for the long haul. I've seen it happen time after time on smaller tracts ranging anywhere from 250 acres on down to the 40 acres mentioned above. It can be done!

Summary — Chapter 3

- High-protein nutrition is the key to any serious trophy management program.
- Through a combined program of planting food plots, supplemental feeding and putting out minerals, protein gaps that occur during certain times of the year can be eliminated. Once this is achieved, your deer will be getting all the protein they need.
- If you are having trouble finding a home for your extra venison from harvesting does, check out your local food banks and "Hunters for the Hungry" programs.
- The "Banks Farm Way" will work just as well on small tracts as it does on large tracts.

Here's another power buck that is a direct by-product of the Banks Farm high-protein planting and feeding program. Club member Carl Wilson shot this big nine-pointer in 1997. The buck grossed 148 B&C points. Although the buck only had a 14-inch inside spread, he had great antler mass. His bases were seven-inches around. This is what protein will do for your bucks!

Chapter 4

High-Protein Food Plots
Building a Protein Factory for Deer

Okay, I admit it. I'm obsessed with high-protein plantings for deer. But I warn you in advance. If you want to make some positive changes on your property and start growing some true power bucks of your own, you'll probably come to the same realization that I did back in the mid-'90s. Sooner or later, you'll begin to realize just how important protein is in a buck's diet, especially as it relates to his "horns."

Once I started learning a little bit about deer nutrition, the word "protein" became all-important. It didn't always use to be that way. As I mentioned in an earlier chapter, there was a time when I thought any green field planted with any summer or winter forage was sufficient to fill the nutritional needs of the deer. It didn't matter what was in that field. It could be wheat, oats or rye. As long as it was green, and as long as the deer appeared to be eating it, everything was hunky-dory.

As time went by on the Banks Farm, however, I started reading everything I could get my hands on about deer nutrition. If I saw a magazine that had an article about nutrition in it, I'd buy the magazine and read the article. I attended Quality Deer Management Association meetings in Georgia and listened to what biologists were saying about deer nutrition. I also talked to other landowners and professionals about their various supplemental programs. If I thought someone else's idea would work on the Banks Farm, I'd try it and see how it would apply in our situation.

It took a while, but we eventually began to realize just how important high-protein plantings were to our overall management program and our ultimate goal of growing exceptional bucks. If I'm a protein fanatic, then so be it, but I firmly believe that being obsessed with growing bigger and better deer has helped us get to the point where we are today.

Filling in the Gaps

Not only is protein the magical ingredient, but in order to have the healthiest deer possible, those deer have to have high-protein foods available to them 365 days a year. As soon as we realized that there were serious gaps during certain times of the year when our deer were getting little to no protein at all, we started taking steps to solve that problem. This is something that many hunting clubs fail to understand. Since it's one of the most important points in the entire book, I'm going to repeat it again and highlight it here. **In order to have the healthiest deer possible, those deer must have high-protein foods available to them 365 days a year. Natural gaps that occur during the year must be filled in. If those gaps are not filled in, your management program will never be complete.**

We had been planting iron and clay peas for several years during the early '90s and they had been very successful to our program. The deer loved them, and they were high in protein. What's more, we killed some super bucks in '94 as a direct result of a great growing season and plenty of rain. But peas are a spring and summer feed. You can't really plant them until late April or the first of May. The most serious gap that we needed to try to fill ran from January to May. We knew we needed to give our deer some good protein earlier in the year so that the bucks could get a jump-start on the antler growing season, and so that the does could get a jump-start on raising healthier fawns.

As mentioned, Dad had been planting a small amount of winter wheat on the farm for years. Wheat is high in carbohydrates, but it doesn't contain much protein. I knew there had to be something else out there that we could plant that would provide the kind of protein we were looking for. As I learned more and more about various plantings, I got very interested in alfalfa. Alfalfa grazing is a year-round perennial, and it's very high in protein. It takes a good bit of maintenance to keep up, but we were willing to put a little "sweat equity" into maintenance if we could get the results we were looking for.

You've got to keep a close eye on alfalfa grazing, and you've got to keep it clean – that is, keep the weevils out by spraying and keep the weeds out as well. Alfalfa grazing is also very drought tolerant and the deer can't kill it by overgrazing. Therefore, I thought it might be just the thing we needed to fill in that early season gap. So I bought some alfalfa grazing seed from Pennington Seed, Inc. and we began using it. It later tested at about 30 percent protein on our farm. I was very happy about that.

About that same time we also started planting some ladino clover. Both alfalfa grazing and ladino clover are great high protein plantings for deer, but as mentioned, ladino clover is not the least bit drought tolerant. We've been under drought conditions for the last six or seven years in Central Georgia and we learned that the hard way. If you plant ladino on the hillsides or other high places and you have a dry spell, it's not going to survive. Ladino is great forage for deer. Its great stuff all the way around – *if* you plant it in the moist bottomlands instead of in the high, well-drained places.

On the other hand, alfalfa grazing is very hardy and drought resistant. It was designed for heavy grazing by cattle, and it holds up well under heavy grazing circumstances. I figured that if it was hardy enough for cattle, it ought to be perfect for our deer. So we started planting alfalfa grazing in a lot of our upland food plots and ladino clover in the bottoms.

I was also interested in trying to find something else that didn't require as much maintenance as alfalfa grazing did, and I started experimenting with yuchi arrowleaf clover. You could plant yuchi in the fall, and like alfalfa, we found it to be very drought tolerant. It's tough, and it produces a lot of forage, also similar to alfalfa grazing. I had talked to different farmers who often used yuchi to over-seed their pastures for cows because it was so high in protein and it was relatively inexpensive to plant. It grows very tall, and it produces a lot tonnage. The deer can't possibly eat it all. That's what you want. And it grows well just about anywhere – particularly in places like old loading dock areas. With a little lime and fertilizer, all you have to do is put it out and it'll do what you want it to do.

We found that the combination of alfalfa grazing and the two kinds of clover went a long way toward filling in those protein gaps that existed from January through May. (In extremely dry years like we experienced during the late '90s, that gap might extend all the way to August.)

If you're a serious trophy hunter, you know that late

winter is often a critical time for bucks. They've just gone through a month or more of intensive rutting activity and they've probably dropped a lot of weight. They've also endured the strain of being hunted for several months, and they may even be carrying battle wounds from previous rutting activity that haven't completely healed. They often start the New Year in early January in a weakened condition. To make matters worse, after they shed their antlers in January or February, most of the natural browse in the woods is gone – particularly most green plants that contain protein. Natural food supplies get awfully short at this time of year.

If we could get the supplemental protein to our deer during those late winter months, we knew it would help them get their systems back up to speed at a time when they needed it the most. Then, by late spring when they started growing their new antlers, they'd already have a little boost. We found that yuchi was a great "gap filler" early in the year. By March or April, the alfalfa grazing was really coming on strong in our upland food plots and our ladino clover was doing well in the bottoms. Collectively, we were beginning to fill in those gaps and have plenty of good feed available whenever our deer needed it.

Yuchi is virtually maintenance free. You simply plant it by itself or mix it in with oats or wheat, and let it grow. I prefer to mix yuchi with oats. For us, that's been a magical combination and I strongly recommend mixing the two together. The two plants complement each other very well. I'll go into more detail about that shortly. You've got to inoculate the yuchi seed to help it germinate, but after that, you're home free. Yuchi is very disease and insect resistant. Most important of all, though, the deer love it. Ladino clover requires more maintenance. Like alfalfa grazing, you have to spray ladino to

This young buck in velvet is almost lost in this field of yuchi arrowleaf clover. This photo was taken in early June 2002 and the yuchi – planted the previous fall – is still thriving. In fact, it's so thick that the oats are buried underneath it. I love food plots like this. The deer can't possibly eat all the clover in this field.

help keep the weeds out because it doesn't grow as tall as yuchi.

Once we started filling in the protein gaps, we started seeing some significant changes in our deer herd. By having all this protein available to our bucks 365 days a year, we can now look back and say with reasonable certainty that our bucks have gained an average of 15- to 20-inches in antler growth during the past 10 years. We have numerous videos of bucks feeding out in our alfalfa fields. One video shows 27 deer in a field and 18 or 19 of those deer are bucks. About two years ago, my cousin Danny Keever shot some video footage of 18 bucks feeding together in one three-acre field of waist-high yuchi clover. It was an awesome sight to watch that many bucks together and see how they reacted to the clover.

They absolutely loved it! They came out of the woods with their heads down, and they continued eating the whole time they were in the field. They fed until they were full. Then they turned around and walked back into the woods. In 20 minutes they had eaten their fill and were gone.

When you've got quality food planted for your deer, they're going to be in it at some point in time during almost any 24-hour period. But beware of what you plant, and make sure the deer like it. A number of hunters I've talked to have planted some of those highly publicized deer plantings that are supposed to be so great. Time and again, these people have witnessed their deer come out of the woods and walk right through one of their food plots without even stopping.

The Importance of Good Records

After 12 years of intensive management on the Banks Farm, the payoff for us has been a substantial increase in both body and antler sizes of our bucks. I know I'm repeating myself, but we attribute this increase directly to the combination of high-protein plantings, supplemental feeding and good mineral supplements that are available to our deer on a year-round basis. One of the areas where we've been a little weak since the beginning of our program has been in the record-keeping department. Back in 1990 when we started our program, we didn't keep too many written records, but we knew what kinds of weights we were seeing on both our bucks and does from year to year and we were aware of any changes. Fortunately for us, those changes came about in the form of noticeable increases.

If you operate a hunting club, you should be keeping some type of records. You don't need to do anything too elaborate, but it's a good idea to get a notebook or note cards and record the body weights of every buck and doe harvested. We've always done a much better job of keeping weight and age records of our bucks than we have with our does. One reason for this is the fact that we shoot so many does on our property it's extremely hard to keep records of every single animal harvested.

During the 10 year period from 1990 to 2000, we saw an average increase in the body weights of our does of about 25 pounds. In 1990, most of our does weighed around 100- to 110-pounds live weight. A heavy doe back then may have weighed in the 120s. By the end of the '90s, we were consistently seeing does that weighed-in the 130s and 140s. We recorded several does that weighed over 150-pounds live weight. We also noticed an increase in the size of our bucks. In the late '80s and early '90s, we obviously weren't shooting that many older-age-class bucks, but some of the best 2 1/2 and 3 1/2 year old bucks we killed weighed-in at around 180- to 185-pounds. In those days, we seldom saw a buck that weighed over 200 pounds. Now, most of our mature bucks are easily topping the 200-pound mark. My Boone and Crockett buck taken in 2001 weighed 232 pounds (live weight). We've taken several others "power bucks" off the farm in recent years in the 220- to 230-pound range. You'll never be able to convince me that protein didn't have something to do with those weight increases!

Weights are important, but the real bonus for us has been the gradual increase in antler mass and tine length on Banks Farm bucks. I can't emphasize this point enough. Even if you hunt in an area where average body and antler sizes are much smaller than what we are blessed to have in Morgan County, Georgia, you definitely can increase the size of your bucks' antlers with the right kind of program. Again, keep your records simple but make sure you have those records on file so that you'll have a road map to compare the progress you are making from year to year.

How Much Land Should Be in Food Plots?

I've read many articles on the subject and I've talked to a number of biologists in regard to this question. Most biologists I've talked to believe you should try to have around 10 percent of your total land area in food plots. If it's any consolation, we've never had anywhere close to a 10 percent ratio on the Banks Farm. As of 2001, we probably had about 100 acres in food plots. That's only about three percent of our total land area if my math is correct. It would be better to have more, but you've got to

Everything we do for the deer on the Banks Farm in the way of food plots and controlled burning also benefits the turkeys. Our spring turkey season in 2002 was exceptional. Here are five happy hunters with three nice longbeards taken on the Banks Farm. On the day these gobblers were killed, Sam Klement and Ron Jones of Spectrum Outdoors were filming a video. (Left to right) Sam Klement, Danny Keever, Jeff Banks, Billy Young and Ron Jones.

We start planting our iron and clay peas as early as possible in May. Here, Horace Wade, my right-hand man on the farm, is getting a food plot ready to plant.

work with what you have.

Like a lot of hunting clubs, we've been limited with the total number of open acres that could be planted. However, we've been able to work around that problem through supplemental feeding and putting out minerals. What's more, our 100 acres of high-protein food plots are evenly distributed around the farm. Our deer never have to go far to find a food plot, and in my opinion, they get every bit of protein they can possibly use.

If you lease a tract that is mostly in timber like many hunting clubs do, it may be very difficult to have even three percent of your land area in food plots. However, you can often find old loading dock areas that get partial sunlight, or old logging roads, or open gas and power line easements that run through the property. These are excellent locations for planting food plots. Most timber companies and landowners are more than happy to work with you when they know you are doing things to help improve their land.

If you're making an effort to keep your deer numbers down by shooting enough does, and if your management program includes a combination of high-protein plantings, supplemental feeding and putting out minerals, you can overcome the lack of open acres for food plots and you can do wonders for your deer herd.

Fall Plantings – the Banks Farm Way
Yuchi Arrowleaf Clover (combined with oats) – (September and October)

Mix about 3 1/2 to 4 bushels of oats with about 12 pounds of Yuchi clover per acre. If you're using a whirlybird on a seeder, mix your oats in with a bag of yuchi (yuchi normally comes in 50-pound bags), put your inoculant on the seed (mix it with water or whatever works best for you), and put it in your hopper. Mix the oats and the yuchi up real well, and you're ready to put it out. Lightly drag it in while you are planting, and you should be set. That's all there is to it. I suggest that you also lime and fertilize as needed.

If we can just get some good rain, these iron and clay peas will come on strong by mid-summer!

We like to plant as early as possible in the fall. I do my best to have our oats and yuchi in the ground, weather permitting, by the first or second week in September. The reason for this is because we like to have our food plots up and running by mid-November. That way, the plants have nearly half of September and the entire month of October to get established and start attracting deer. When the peak of rut arrives in early to mid-November, it gives us a great opportunity to get a good look at some of our bucks when they come out into the fields to feed.

We've very successful in both seeing and killing big bucks out of food plots during the rut in November. In Chapter 2, I mentioned Ken Brown's 160-class buck that was taken in a food plot in 1994. Ken's buck was chasing a doe and they both ran right out across the food plot. I also mentioned Jarrod Brannen's 150-class "milestone" buck taken in '93. Jarrod's buck was a milestone for our club because it was the first 150-class buck taken under our modern trophy management program. It was the 19th deer to come out into the field that day, and the ninth buck. In addition to killing some nice bucks out of our food plots, I can't begin to tell you how many trophy-class deer have been captured on video while they were feeding.

I wouldn't recommend using less than 10 pounds of yuchi per acre, or any more than 15 pounds per acre. You can mix yuchi in with wheat but I wouldn't suggest doing it with rye. Rye tends to come up a lot thicker and there's too much competition for the clover. When the oats pop up and sprout, the yuchi tends to fill in the gaps between the new oat sprouts. After that, both plants grow together extremely well. We've had tremendous success with this combination. The deer absolutely love it. If you get adequate rainfall, the yuchi will get waist high, and you'll see plenty of healthy deer in your food plots.

As mentioned, Yuchi will grow just about anywhere. That's why I'm so high on it. It'll grow in low areas as well as on hillsides. It'll also grow in new loading docks areas. It's adaptable to just about any soil types, and it's hardy. It's high in protein – we've had it tested at 38.1 percent protein. The TDN (total digestible nutrients) tested at 89 percent. All those numbers are very high.

Yuchi can be planted just about anywhere on your deer lease and you should see some good results. You certainly don't have to mix it with oats, but I strongly recommend it. The oats give the deer something to eat while the clover is coming along. Yuchi will usually last until the middle of June. Again, depending on the weather, I may begin to plow up some of our yuchi in late April or early May and put some of those areas in iron and clay peas. The weather has been very tough on us during the past few years. It's been extremely dry, and we've been blessed to have yuchi clover in our food plots. We have had some situations in recent years where we've planted iron and clay peas, but due to a combination of extremely dry weather and the deer eating them, the peas didn't make it.

Ladino Clover – (September and October)

I recommend planting about 25 pounds per acre of ladino clover in any bottomland area that gets adequate sunlight. A little shade is fine, too. The reason I plant nearly twice as much ladino seed per acre as yuchi is because ladino does not get as tall as yuchi, and I like to get my ladino a little thicker. That way it'll establish itself well, and it'll be thick enough to help keep the weeds out. Also with a good thick stand of ladino in the bottoms, if it does get dry and some of the plants die,

Horace Wade and I are getting ready to do some serious planting.

you'll have a thick enough stand so that some of it will survive. I don't recommend planting ladino clover on any type of hillside or hilltop. The root system is simply too shallow. If drought conditions occur, the plant will stress out and die. Lime your soil well, and fertilize as needed.

I plant all our ladino clover in the bottomland at the same time I plant yuchi, in early September. You can plant ladino right up to the first of November, but I like to get it in the ground early. That way, if something happens, we still have time to correct the problem before cold weather sets in.

Other Clovers –

On the Banks Farm, we've more or less found our niche with yuchi arrowleaf and ladino clover. We've never done much with the red clovers, or the crimsons, or any of the other brands. That's not to say that the other brands are not excellent clovers. For the money, however, and the results we've been getting, we've been very pleased with the two types of clovers we use.

If you don't own your own equipment, hiring out the work can usually be arranged for a reasonable price.

Alfalfa Grazing – (September and October)

I like to plant about 25 pounds of alfalfa grazing per acre in early September because I prefer a good, thick stand and it helps keep the weeds out. Alfalfa grazing needs to be planted on high, well-drained ground. It does not do well in low, moist areas because it tends to get a fungus that will kill it. Alfalfa grazing is very drought tolerant. If planted in the right areas, it should thrive. The reason I started planting alfalfa grazing is because it fills in one of those gaps I was talking about earlier. In the spring, we can't get our iron and clay peas in the ground much before the first of May. It'll take another two months after that before they become good grazing material for the deer, and by then, mid-summer has arrived.

To help fill that gap between January and June or July, we started using alfalfa grazing. It's high in protein, and it doesn't start losing its value until mid-summer. Yuchi is planted in the fall, and it provides great protein for the deer from January through May. That really helps them out early in the year and also during the period when their antlers are just beginning to grow.

We'll leave some of our food plots in yuchi, and take another food plot not far from it and plant it in iron and clay peas so that the peas will be coming on strong just as the yuchi is starting to die off by the end of June or the first of July. During extremely dry years like we've experienced since 1996, the alfalfa also helps fill in that protein gap during the warmer months if the peas are not doing well.

If all else fails and it's so dry that none of our food plots really produce the quality of feed that we hope they will (and we've had some seasons like that during the past few "drought" years), the supplemental feeders should take over and give the deer some protein when they really need it. In our experience, the drier it gets out in the woods, the more the deer are going to hit those feeders.

I know I keep harping on shooting does, but if your doe population is kept down to where it should be, you should also have a lot of natural browse out in the woods that the deer can eat during extremely dry periods. On the other hand, if you're plagued with too many deer on your lease or hunting club, you won't see much natural browse, especially those plants that the deer favor. They simply don't exist on an over-grazed range.

The key to growing alfalfa grazing and clover successfully is in preparing the site. You need to do a good job in plowing so that you have a very good seed bed. Once you get your seed out, drag the site lightly with a chain or other device. If you plant your seed too deep, it will not come up. Also the site needs to be properly limed and fertilized.

Spring Plantings – the Banks Farm Way
Iron and Clay Peas – (May and June)

We usually plant about two bushels per acre of iron and clay peas around the first week in May. We plant them by themselves in individual food plots. From time to time, I occasionally mix in a little millet for the turkeys. Peas are great as summer forage for deer. They're good right up until the first frost in late October or early November. They're high in protein, and they produce a lot of tonnage. Deer love them. We've learned that peas are also a great way to gauge your total deer population. If the deer seem to be eating up all the peas right out of the ground as soon as they pop up, that may be a good indication that you're not shooting enough does. On the other hand if the deer are *not* eating up all your peas, your total population is probably closer to where it should be.

Turkey Plantings –

We've never planted any chufa specifically for turkeys but it is a great supplement. What we have been doing is planting millet up and down the hedgerows for both the turkeys and the quail. Once planted, we leave the millet standing. We also try to leave a lot of standing cover around the edges of our pea fields and other food plots for the turkeys. They like to nest in the high stuff along the edges of the fields and what we leave standing also provides good feed for them in the fall. That has worked out real well.

Our turkey hatches have been unusually good in recent years, and I think our deer program has complimented the turkey population. Turkeys love to eat the seeds from just about any kind of grass, and I've seen their craws absolutely stuffed with oats. They also like clover. It's not uncommon to see a herd of deer and a large flock of turkeys feeding together in one of our food plots. We have a lot of turkeys on our property and the flocks seem to be in great shape. Burning has also helped our turkeys. Opening up the woods through burning not only provides easy access to insects and seeds on the ground, but the visibility factor no doubt helps protect our turkeys from predators like foxes and bobcats.

Generally speaking, we've found that just about anything we do for the deer on the Banks Farm also benefits the turkeys.

Other Plantings –

Hairy vetch is an excellent plant. We've never planted a lot of vetch on the Banks Farm because we have a lot of natural vetch that comes up in our fields. In the old days, farmers used to use it for hay on our farm, and there is still a lot of it around. It's very high in protein and it's a great food source for deer. The main reason we don't plant any vetch now is because we have our hands full with other plantings. Furthermore, we've more or less found our niche with yuchi arrowleaf and yuchi has really served us well.

We don't plant any rye, either. Rye tends to come up much thicker than either wheat or oats, and it tends to choke out your clovers regardless of what kind you are using. Therefore, we prefer to use either wheat or oats in our mixtures instead of rye. Both are fine to mix with clover, but as I've mentioned before, the combination of yuchi arrowleaf and oats has been a magical combination for us.

In the past, we've also planted grain sorghum in some of our food plots, but the deer will really mow it down if you don't have enough of it. We've also planted corn from time to time, but you have to pray for rain if you plant corn, and we've had a real shortage of rain during the past five or six years. Because of the rain factor, we've pretty much focused all our efforts on planting the majority of our food plots in the yuchi arrowleaf and oats combination because been yuchi is not as expensive as some of the other clovers, it's extremely hardy, and it's very high in protein as I've mentioned before (ours has tested at 38.1 percent).

How We Established Our Food Plots on the Banks Farm

On the Banks Farm, we now have a total of about 25 food plots. They range in size from one acre up to 13 acres in size. When we got serious about planting high-protein food plots during the early '90s, we basically took whatever we had to work with and tried to make the best of it. That's all anybody can do. If I had the opportunity to take an area around a creek bottom and clean it up and plant it, I did it. If I had another area on a hillside somewhere that could be cleaned up and planted, I did that also.

As mentioned, we also cleaned up several old fields on the farm that had been allowed to grow up. In one area, we had about five acres of beetle-killed timber. We cleaned that up and made it into a food plot as well. It took a lot of hard work to get some of these areas into good, productive food plots. At first it was kind of hit or miss. But after a while, things started falling into place for us. After having one food plot here and one over there, eventually we ended up with a fairly nice pattern of food plots situated throughout the farm. You can do the same thing.

The reason I mention this is because anyone can do the same thing on any piece of land. I never had the luxury of sitting down and coming up with the "perfect plan" for our food plots. Through a lot of hard work, however, we were able to develop a good system of food plots that have served us well. More importantly, they've served the deer well. Take what you have and make the most of it. If you are limited with little to no area available for planting food plots, supplemental feeding and putting out minerals become more important than ever.

Lime and Fertilizer –

It doesn't really matter whether you lime in the spring or fall. The important thing is to make sure that the pH level of your soil is well-balanced so that every plant in the food plot absorbs the maximum amount of fertilizer you put out. If you're not sure about your soil, take some soil samples and send them to your county extension agent. They'll tell you what your pH level is, what it should be, and whether or not you need to put out additional lime. New food plots in areas that were previously wooded like loading dock areas generally require a lot more lime because the soil is much more acidic.

If you lime correctly, you'll seldom need to re-lime your food plots more than once every three or four years. We normally put out two tons of lime per acre in our food plots every four years. We have our soil tested from time to time to make sure it is balanced.

Once our fields are properly prepared and ready for planting (that is, plowed, harrowed and smoothed up – especially if you're planting on new ground), what we normally do is lime those fields and fertilize at the same time. Then we plant our seed right on top of the lime and fertilizer. If you have a club, and if you're trying to get everything done during a designated "work week," this method works extremely well. It's also fine to plant your food plots and fertilize at a later time.

On the Banks Farm, we normally fertilize our fall food plots after the plants are established and growing. As a rule, we fertilize one time each year and we don't put out additional fertilizer until the following year. We normally fertilize at the rate of about 300 to 400 pounds per acre. Once your food plots are up and growing, if some of the plants start to yellow a bit you might need to add some light fertilizer about mid-way through the growing season. With new food plots especially, it might take about six months for the lime to really kick in good, and it might be necessary to add some additional fertilizer to help get that food plot established. That certainly won't hurt anything.

Summary — Chapter 4

- **To reach and maintain optimum growth levels, deer need a diet that contains 16- to 17-percent protein 365 days a year. Even with some planting and supplemental feeding, protein "gaps" can occur at certain times of the year. Protein gaps most often occur from January through May, but they can also occur in late summer. In order to have a successful deer management program, all gaps must be filled.**
- **Protein gaps can be filled whenever they occur through the right combination of planting, supplemental feeding, putting out minerals and making sure that natural foods in the woods are not over-browsed.**

Former Banks Farm club member Bruce Coleson took this thick-antlered 5x5 in 1997. The big buck grossed 140 B&C points. It was obvious Bruce's buck was getting plenty of protein in order to grow a set of antlers like this. Supplemental feeding is a must for any serious trophy management program.

Chapter 5

Supplemental Feeding

I've already told the story about Dad putting out home-made feeders in the early '80s. Instead of putting out corn like most people do, Dad wanted to offer our deer something with a little more nutritional value than plain corn. (Corn is very low in protein and very high in carbohydrates. It's fine to put corn out as long as you have plenty of other high-protein supplements to go along with it, and the deer love it.)

Dad couldn't find any type of deer feed locally because there was not much demand for it in the South. So he ordered a special blend of pellets from Purina Mills, and had them shipped all the way to Madison from Arkansas. Unfortunately, his supplemental feeding program never really got off the ground and it kind of fizzled out. But it was a beginning, and today we know a lot more about feeding deer than we knew in 1980.

In 1995, when I built a feeder to put in my front yard so that we could watch deer from the house, I started buying all our deer feed from Godfrey Feed and Seed in Madison, Georgia. As mentioned in Chapter 3, we still use Godfrey's blend today. It's essentially a sweet feed blend very high in protein. The deer love it. It consists of

When Dad built some of our first home-made feeders during the early '80s, he inadvertently put dividers in them to make them stronger. He never thought about the fact that antlered bucks could not get their heads down in the feeders to eat. We all learn from our mistakes!

rolled corn and pellets containing molasses and other minerals and vitamins. Most experts will tell you that deer need to receive 16- to 17-percent protein in their diet in order for them to reach maximum growth potential. Godfrey's feed contains about 16- to 17-percent protein so the deer get everything they need from it. It's a great supplement.

After I killed my record buck in October of 2001, Godfrey's Feed and Seed started bagging a new blend they named "17-Point Whitetail Deer Feed" in honor of my buck. (My buck had obviously been eating their feed. We had plenty of trail camera photos to prove it. Actually, my buck ended up being officially measured as a 16-pointer, but as of the time he was killed, we thought he was going to have 17 scorable points.) This new blend has a slightly higher protein content than their older blend.

Supplemental Feeding

After seeing how the deer responded to that first feeder I put out in '95, I started building and putting out several more feeders each year around the farm. Today we have approximately 15 feeders located throughout the property or one feeder for every 200 acres. The type of feeder we build is a basic trough feeder covered by a good roof. Most of our feeders stand 14 to 18 inches off the ground so the deer can get to them easily. Our feeders are open in the middle and there is plenty of clearance between the trough and the roof so that a buck with large antlers can eat with no problem.

I mentioned the fact that some of Dad's early feeders had wooden dividers in them. They were fine for does and fawns, but a buck with antlers could not get his head down in the trough to feed. We laugh about it now, but that's just one of the many mistakes we made back in the

Our new feeders are well off the ground and have plenty of "head room" for a buck with a big set of antlers. In times of extreme dry weather like we've experienced during the past four or five summers, our deer clean these feeders out at an incredible rate. After I killed my record buck in 2001, Godfrey's Feed and Seed in Madison, Georgia, started bagging a new blend of feed they named "17-Point Whitetail Deer Feed" in honor of my buck. It contains about 17-percent protein and the deer love it!

early days. Trial and error has taught us a lot.

Our trough feeders are about five feet long and hold between 400 and 500 pounds of feed when full. We've never used any automatic or broadcast-type feeders. I've never liked the idea of having any feed hit the ground. It's too expensive and too much gets wasted. Also, with so many deer using the feeders, it can easily get contaminated if it's on the ground.

We try to make sure that our deer get clean, uncontaminated feed at all times. I'm a fanatic about making sure the feed is clean. Between the birds, the chipmunks, the 'coons and the 'possums, we're going to lose a certain amount of our feed to the critters, but I think it's much healthier for the deer to eat out of troughs. Speaking of 'coons and 'possums, we try to avoid putting our feeders near creeks because of the high 'coon traffic in those areas. We try to place our feeders in strategic locations in the woods where the deer can easily find them during their daily travels. The 'coons still find them, too, but our losses are not nearly as bad as they would be if the feeders were closer to the creeks.

You can get away with putting straight corn out on the ground, but you can't possibly put pellets out on the ground because they tend to dissolve as soon as they get wet and they mildew very quickly. In fact, with the least amount of rain, unprotected pellets tend to get moldy right away. If any of our pellets in the troughs do get wet from a heavy rain, we usually shovel them out right away and scatter them in the woods somewhere for the critters. We like to keep fresh, dry feed in our feeders at all times.

We feed all the way up to late August. We usually start feeding the day after hunting season closes in early January. From January through late August we probably fill our feeders about 10 times, or roughly once every

My dad Lamar Banks with yet another high-protein "power buck" taken off the Banks Farm. This one, taken in '98, grossed 148 points. The buck's rack was a main-framed 4x4 with heavy mass and several stickers. There's no question that a high-protein diet contributed to this buck's exceptional rack.

three weeks. In late May 2002, while this chapter was being written, we put out 2,600 pounds of feed. The deer were in those feeders immediately. They really seem to hit the feeders hard in later spring and early summer. As soon as the peas start coming up good in late June, they begin to lay-off somewhat.

Obviously it's against the law to hunt over bait in many states, and we usually stop feeding well before deer season opens. I'm a fanatic about making sure we obey all the game laws, and I make sure all our feeders are completely cleaned out before we start hunting. A few of our feeders are placed in the edge of the woods near food plots. That way, if someone wants to watch a food plot during late summer before deer season opens, they can also watch the feeder at the same time. Sometimes they'll glimpse a big buck at one of the feeders. If they are lucky, they might get him on video. This is a great way to "preview" some of our bucks for the fall hunting season.

No matter where you place your feeders, it's important to have them in areas that are easily accessible by truck. Most of our feeders are portable. If necessary, they can be loaded onto a small trailer and moved from one location to another. In fact, just to be on the safe side and avoid contamination, we like to move our feeders on a regular basis. Sometimes we only move them 40 or 50 feet away from where they were set up, just to give that area a rest.

We've been putting out deer feed for the past six or seven years, now. Along with our minerals supplements, I think it's really made a difference in our deer herd — especially during the dry times we've experienced in recent years. Since planting food plots and supplemental feeding has become so popular with whitetail hunters in recent years, many good companies are now marketing some excellent blends of deer feed. Check with your local feed store to find the right blend for your management program.

Summary — Chapter 5

- **A high-protein supplemental feed can help give your deer much needed protein supplements during certain times of the year when they really need it, namely from January through the end of August.**

Mentioned earlier in the book, here's another picture of Danny Keever with his beautiful 10-pointer that I refer to as a "genetically superior freak." Although the buck's rack grossed in the low 130s, he field-dressed at an unbelievable 101-pounds. We believe the buck might have been only 2 1/2 years old. He certainly looked like a mature buck and that's why Danny decided to take him. Could mineral supplements and other high-protein feedings have anything to do with his trophy rack?

Chapter 6

Mineral Supplements

I've already mentioned the importance of mineral supplements in Chapter 2 and Chapter 3. While I was still attending Georgia Southern University in Statesboro in 1986, I stumbled across a salt-free mineral supplement in a local feed store. If my memory serves me correctly, I think the salesman told me it was designed for domestic hogs. (Obviously hogs can't have a lot of salt in their diet.) At any rate, the minerals contained about 30 percent calcium, about nine percent phosphorous plus a lot of other important minerals. Because of the high calcium and phosphorous content, I knew it would be good for our deer. So I bought a ton of it and brought it home.

Along with my brother Lane and my good friend Pat McDevitt, who played football with me at Georgia Southern, we went around the farm one weekend and put it out. We made several different mineral licks. Since that particular mineral supplement contained no salt in it, I went out and bought some trace minerals locally that contained about 95 percent salt. We mixed in about 15- to 20-pounds of the salt into each mineral lick we had made so that the deer would be attracted to them. Then we left them alone.

Banks Farm club member Danny Keever makes a new mineral lick. We use a special blend of minerals bagged by Godfrey's Feed and Seed in Madison, Georgia.

To my amazement, the deer absolutely tore those mineral licks up. The warmer it got, the harder the deer seemed to hit those licks. That experience convinced me that there was something in those minerals the deer really needed. As time went by, I had a hard time getting those same minerals in Statesboro. I think the company supplying them went out of business. So we started buying other mineral supplements wherever we could find them.

Eventually I went up to Godfrey's Feed and Seed in Madison and talked to Weyman Hunt, Charles Hendrix' first cousin, about mixing up a special blend of minerals specifically for the Banks Farm. Together we came up with a granular mineral supplement that contained about 20 percent salt (which is important for attracting the deer).

Today we use the Godfrey Feed and Seed blend on a regular basis. It has a good calcium and phosphorous content which are both important minerals for growing strong bones and antlers. It also contains all the other important minerals which are so vital for growing healthy deer – does, fawns and bucks.

Making a Simple Mineral Lick

Just like with our supplemental feed, we like to get our minerals out as soon as deer season is over, but no later than February or early March. That way, minerals are available to our deer in late winter when they need them the most. We'll find a good spot in the woods that is easily accessible by truck. Ideally, that spot will have a lot of deer activity around it. I like to find a spot that is

When a mineral lick is torn up like this one, you know there is something in it the deer crave.

primarily in the shade most of the day so the deer can pick on them during daylight hours in the summer when it's hot if they want to. I also try to find a spot that is close to a timber change.

I'll make an opening in the leaves about three feet wide with a rake. Depending on the terrain, I might dig out a small impression in the soil as well. Then I'll pour out two 50-pound bags of minerals into a large pile. That's it. I don't mix it in with the soil or anything else. The deer do the rest. They visit all our mineral licks on a regular basis.

I've been putting out one ton of minerals on our property every year since 1986. I know our mineral licks help the mothers and babies in the spring, and I know they help our bucks. Currently we have about one mineral lick for every 150 acres on the Banks Farm. We maintain close to 20 licks in all, and it's not a big expense. Each year, we put out two 50-pound bags per lick. That equals one ton. In some places, the deer absolutely annihilate certain licks. We try to keep an eye on those licks to see how fast the minerals disappear. If necessary, we'll "freshen up" these licks during the year by adding another 100 pounds or so in late spring or early summer.

The deer know where all our mineral licks are located, and they visit them whenever they need to as they cruise around during their normal daily routines. The minerals are always available during the antler growing season as well as when the mothers are nursing their fawns. By the end of the summer, the licks are usually pretty well used up. That's good, because we don't want to be accused of hunting over bait when archery season starts in September. Once deer season is over, it's time to put out another ton of minerals.

Summary — Chapter 6

• Mineral supplements are very important to help round out your overall supplemental feeding program. Mineral supplements should be available to your deer on a continuous basis from early January until late summer.

Compared to other yearly costs of maintaining a trophy deer management program, the costs of planting food plots and putting out supplemental feed and minerals is a real bargain – especially if you can grow true "power bucks" like this one on a consistent basis. Here, my twin brothers Loy and Lee pose with my 161 2/8 B&C point trophy buck taken in 1994.

Chapter 7

Putting It All Together –
The Low Cost of Planting and
Supplemental Feeding

By now, you may be getting a little nervous about the cost of planting food plots, and putting out supplemental feed and minerals. Before you decide to take out a second mortgage on your house or sell your children, read on. The cost of these programs can be a lot less than you think — particularly if you split it up between eight or ten club members. In fact, I've often said that a box of high-powered rifle bullets frequently costs almost as much as 50-pound bag of yuchi arrowleaf clover seed.

Also keep in mind that we didn't start off doing all the things we're doing today. It was a very gradual process to get to the point where we are at. But even now with the fairly large operation that we maintain, our management program runs smoothly because everyone pitches and goes the extra mile. We decided a long time ago that we could have a mediocre management program, or we could strive for excellence. Everyone in our club decided to take the high road and try to make our club the best it could be. Another thing I've often said in

I took this thick-antlered 10-pointer in 1996. He grossed just under 150 B&C points and netted 144. He weighed 224 pounds (live weight). This buck was the product of a lot of hard work and commitment – what we call "sweat equity" on the Banks Farm. I believe the cost of a top-notch trophy management program on any lease or hunting club in the South can be kept to a minimum, if you and your members are willing to invest a little sweat equity.

recent times is that I could leave the farm for an extended period of time and our program would still run smoothly in spite of my absence because all the guys know what to do and they tackle their various jobs without being asked.

I've talked about "sweat equity" before. I believe mountains can be moved through commitment and hard work. Each hunter in our club has a special talent and a specific job to do that usually draws from that talent (more on that in Chapter 10). On the Banks Farm, we take advantage of that talent to get the job done, but we don't take advantage of the individual club members. The guys in our club do their jobs with to the best of their ability because, like me, they want to be a part of a cutting-edge management team that has the potential to produce some incredible "power" bucks. In 2001, the combined efforts of many years did produce a buck that we're all very proud of. But we think even bigger and better things are in store for us in the future because we have a passion for what we do and we want to see the best possible results from our hard work.

The reason I keep mentioning this is because you and your club members can do a lot of the work yourselves and save a lot of money. But even if you have to hire some of the work out, you might be surprised at how reasonable these various supplemental programs can be. During the past few years, I've talked to several different hunting club presidents who are afraid to take on a program like ours because they think the costs will be astronomical. Let me say two things here. 1) The cost of a top-notch program can be kept down to a reasonable level, and 2) If you don't have a program that covers all the bases 365 days a year, you'll never fully reap the rewards of your efforts.

I believe a lot of deer management programs never reach their full potential because the participating members are not covering all the bases. You can't do 25 percent of the job and expect to have a successful program. You can't do 85 percent of the job. If your goal is to have a top-notch trophy management program aimed at growing some true power bucks, all the pieces have to work hand-in-hand so that 100 percent of the puzzle fits together. In order to do that, you've got to put in 110 percent effort. It's that simple. No one thing works by itself to produce big deer on the Banks Farm. Collectively, however, the combination of all of the different things we do builds a tremendous year-round nutritional program for our deer.

Let's take a closer look at what it costs us for seed, deer feed, and mineral supplements on a yearly basis. I realize that prices vary from area to area. Prices of certain items can also vary depending on availability and market demand. Since we do much of our own work, most of these costs do not include labor. However, because the Banks Farm management program has evolved into a fairly large operation, there are instances where it is much more economical for us to buy in bulk quantity and pay others to have certain jobs done.

For instance, we buy all our lime and fertilizer from Piedmont Ag Service in Madison, Georgia and the price we pay includes spreading. When you have multiple food plots ranging from 2- to 12-acres in size like we do, it's cheaper to pay someone like Piedmont Ag to come in and perform these services instead of trying to do it ourselves. They are very efficient and they get the job done very quickly. In some cases, they also apply weed killer. On smaller food plots ranging from one-half acre to a few acres in size, we normally do all the work ourselves.

Approximate Annual Costs on the Banks Farm

Seed –
- Iron and clay peas — $16.55 per bushel
    (60 lb. bag. Prices do fluctuate.)
    Total acres on Banks Farm –
    70 acres (on rotating basis) = 140 bushels
    Total cost: iron and clay peas –
    $16.55 x 140 bushels = $2317.00

- Yuchi arrowleaf clover and oats mix –
    yuchi clover – $70 per 50 lb. bag;
    oats – $6 per bushel
    Total acres on Banks Farm –
    60 acres (on rotating basis) =
    12 lbs. of yuchi clover per acre;
    4 bushels oats per acre
    Total cost: yuchi clover –
    $16.80 x 60 acres = $1008.00
  Total cost: oats – $24 x 60 acres = $1440.00
                                     $2448.00

Some trophy management programs never reach their full potential because the participating club members are not covering all the bases. You can't do 75 percent of the job and expect to have a successful program. You've got to go all the way with high-protein food plots, supplemental feeding and putting out minerals.

- Ladino clover – $140 per 50 lb. bag
  Total acres on Banks Farm –
  10 acres @ 25 lbs. ladino clover seed per acre
  Total cost: ladino clover –
  $70 x 10 acres = $700.00

- Alfalfa Grazing — $130 - $149 per 50 lb. bag
  Total acres on Banks Farm –
  20 acres @ 25 lbs. per acre
  Total cost: alfalfa grazing –
  $65.00 x 20 acres = $1,300.00

Lime, Fertilizer and Weed Killer –

As mentioned, we recommend putting out about two tons of lime per acre every three or four years (as soil tests dictate). We normally fertilize our food plots at a minimum of 200 pounds per acre. We recommend a good 10-10-10 or a good 5-10-15. We buy our lime and fertilizer in bulk quantity and have it spread by Piedmont Ag Service.

It takes a lot of protein to grow a heavy-weight like this. I firmly believe that anyone in the Southeast can increase the size of their buck's antlers by at least 20 percent with the right kind of nutritional program.

Lime –
 Total acres in food plots on Banks Farm:
 approximately 100 acres
 Total cost: $35 per ton (includes spreading) x 200 tons
 Two tons per acre every four years = $7000
 Total annual cost:
 ($7000 divided by four years) = $1750

Fertilizer –
 Total acres in food plots on Banks Farm:
 approximately 100 acres
 Cost: 50 lb. bag (5-10-15) @ $5.50 per bag
 x 4 bags per acre = $22 per acre
 Total cost: $22 per acre x 100 acres = $2200

Spraying for weeds –
Alfalfa grazing and ladino clover
Total cost:
approximately $20 per acre x 30 acres = $600

Supplemental Deer Feed –
Godfrey's 17-point Whitetail Deer Feed –
$7.00 per 50 lb. bag or $280 per ton
Total yearly cost: 56 bags (or 2800 lbs.)
x $7.00 per bag (10 times per year) = $3920

Mineral Supplements –
Godfrey's Deer Minerals – $13.50 per 50 lb. bag
Total yearly cost: 40 bags per year x $13.50 = $540

Total Annual Costs of Planting and Supplemental Feeding on Banks Farm

Maintaining 100 acres in Food Plots =	$11,315
Supplemental Feed	$3,920
Minerals	$540
Total Costs	**$15,775**

All of these figures are approximate. The prices of seed, fertilizer and other costs can fluctuate widely, and other variables can affect yearly costs. With our supplemental feeding program, for instance, the deer eat a lot more feed during extremely dry years. Since the last four or five years have been unseasonably dry in Central Georgia, we've been putting out supplemental feed at very high levels.

I realize we are talking about a sizable amount of money here. But please keep in mind that our program on the Banks Farm covers 3,000 total acres with around 100 acres in food plots. Maintaining high-protein food plots and putting out supplemental feed and minerals are three crucial elements for any successful trophy management program. And like I said at the beginning of this chapter, if you divide theses costs up among 10 or 12 dedicated hunters, it's a lot easier to absorb the costs.

I believe that anyone in the Southeast can do the same thing we're doing and make substantial gains with growing bigger bucks on tracts as small as 50- to 100 acres – as long as the five keys in Chapter 2 are put into practice. On a smaller tract, the cost of maintaining a few acres in food plots along with putting out supplemental feed and minerals will be considerably less. Obviously, you can't have 10 or 12 hunters on a tract of that size, but you don't need that many hunters because your costs are much lower.

Each year, hundreds of hunters from across the Southeast pay anywhere from $2,000 to $4,000 or more to travel to places like South Texas, Montana or Canada for a one-time chance to kill a "trophy" buck. In the vast majority of the cases I've seen, if a hunter does bring back a buck, that "trophy" buck will do well to gross 135 B&C points. I'm not saying that bigger bucks are never killed. Record bucks are killed each year by out-of-state-hunters in all of those places. Furthermore, I'm not in any way trying to put down the idea of traveling to another state for a whitetail "dream" hunt. I love to hunt whitetails as much as anyone else. If I get a chance to go on a good hunt somewhere, I'll go in a New York minute.

What I am saying is this: Why not take all or part of that same money and invest it in your trophy deer lease on food plots and supplemental feeding? Instead of a one-time hunt to another state, you'll have many opportunities to hunt on your deer lease in the future. Also, instead of going to Texas or somewhere else and settling for a buck that might score 135-points, why not build-up your own property at home through high-protein food plots and supplemental feeding so that you have a real chance of killing a 140- to 150-class buck or larger a few years down the road? You *can* do it at home. That's the point I'm trying to make.

Summary — Chapter 7

• **Compared to all the other yearly costs of running and maintaining a trophy deer management program, the cost of food plots and putting out supplemental feed and minerals is a real bargain – especially when you consider the benefits.**

Maintaining good natural areas on your deer lease is so important to the health of your wildlife. We've found that almost everything we do for our deer benefits the turkey population as well. Here, Carl Wilson's 13-year-old daughter Carlynn poses with her first turkey ever – taken with a 20 gauge shotgun in the spring of 2002. Carlynn's gobbler sported an 11 3/16 inch beard and weighed 20 1/2 pounds. It had no spurs.

Chapter 8

Other Management Considerations

Natural Browse for Deer and Turkeys

Natural browse is critical to the health of the deer herd. No matter how much you feed your deer through supplemental programs, they still need a wide assortment of natural browse in their diet. The only way to have good natural browse on your deer lease is to keep your deer numbers down to a reasonable level. If you have too many deer on your property, you can often walk through the woods and see evidence of a browse line. You can also tell that you have an over-population problem when certain favored natural foods like honeysuckle are eaten down to the woody fiber and you see no sign of any new growth whatsoever. By the same token, if you've been doing a good job of keeping your deer population under control, you should see an abundance of natural foods out on the woods. Favored plants like honeysuckle will have a lot of new growth on the new shoots close to the ground.

Some hunters fertilize their honeysuckle patches and their favorite oaks trees every year. We've never fertilized any of our natural browse on the Banks Farm simply because we have our hands full trying to maintain our food plots, but it's not a bad idea. One hunter I read about from New York State has a favorite oak tree that he fertilizes on a regular basis. This hunter claims that the acorns from the fertilized tree are much larger and much tastier to the deer than acorns from other nearby trees that go unfertilized.

I'm sure there is something to this hunter's claim. I know hunters in Georgia who fertilize certain honeysuckle patches every year without fail. These hunters also claim that the local deer prefer the fertilized honeysuckle to unfertilized browse nearby. If you fertilize an oak tree or a honeysuckle plant on a regular basis, it stands to reason that that tree or plant will be healthier, grow larger and produce bigger acorns or new growth than similar trees or plants that do not receive the extra nutrients. It also tastes better and it's more nutritious to the wildlife using it. Steaks from corn-fed beef definitely taste better and contain more protein than steaks from some rangy old cow that has never had the proper nutrition.

You can also prove this point with food plots. If you take a well-maintained, well-fertilized food plot, and compare it to one that hasn't been maintained, the deer will always walk right through the one that hasn't been maintained to get to the one that has. I've seen this happen many times with my own eyes. The well-maintained food plot will contain plants that are more nutritious and better tasting to the deer. Deer have an incredible ability to gravitate toward the very best, most nutritious foods around – whether they are natural foods out in the woods, or supplemental foods put out to help them attain their ultimate potential.

Even though we've never fertilized trees or honeysuckle on the Banks Farm, we have taken several areas

The question of whether or not to take management bucks on your property can be a very difficult question to answer. I shot this little buck in 1992 but I should have let him go. He was 4 1/2 years old and he had 13 points (6x 7). All he needed was some good food!

containing hardwoods and cleaned out most of the poplar and sweet gum trees from those areas. That way, the oaks and hickories remaining in those stands don't have to compete with what we consider to be non-productive trees that have very little value to wildlife.

Some hunting clubs plant oak trees, dogwoods, apple trees or other species beneficial to wildlife on their property for the long term benefits. We've never actually planted any hardwoods on the Banks Farm, although we do manage our pine timber and we have planted pines in several areas as part of our overall farming operation.

Keeping Traffic to a Minimum

The Banks Farm consists of approximately 3,000 acres including a large tract that we lease. Since we maintain about 100 acres in food plots, the vast majority

Charles Hendrix (left) took this eight-point cull buck on Innisfail Farm in October of 2001. The buck weighed 235 pounds (live weight) and was probably 4 1/2 or 5 1/2 years old. This buck definitely needed to go.

of our land is wooded. We have several areas in planted pines as mentioned. We also have many acres of mature pines and hardwoods. We have good farm roads throughout our land that gives us access to just about every corner of the property. Although we hunt nearly all of our woods, we go to considerable trouble to keep human traffic in our woods down to a minimum.

The deer always seem to know when hunting season has arrived. Even though they're used to seeing us working out in the fields and driving all over the property throughout the year, as soon as they get that first whiff of human scent out in the woods in the fall in places where they aren't used to smelling it, they know something is up.

Mature bucks need to feel secure on any piece of property, and too much hunting pressure can cause them to leave their normal range – at least temporarily. We are very fortunate on the Banks Farm to not only have a lot of buck sightings, but we see many of the same bucks over and over again throughout the year. This means that we must be doing something right. Our hunters do not drive all over the property on four-wheelers or go out into the woods and move stands around on a daily basis. Whenever they do go into the woods, they try to be very quiet and very discreet. I think this, plus the fact that our land offers the deer a lot of good food and a lot of good areas to hide, accounts for the fact that the larger bucks seem to move around much more freely and they don't seem to feel a need to leave the property because they are being pressured. This is so important.

Some hunting clubs set aside a block of woods where no one is allowed to set foot during hunting season. The idea is to give the deer – mainly mature bucks – a permanent sanctuary where they will always feel safe. We've never done this on the Banks Farm because we have so many areas where big bucks can go and hide, but it's not a bad idea. It's just one more way that you can manage your property or your lease for maximum benefits.

Controlled Burning as a Management Tool

My dad started doing some controlled burning on the farm years ago as a way to clean out some of the thick areas and help generate some of the new browse. After I got out of college and we initiated a serious management program in 1990, we started doing a lot of burning on the farm. I had several friends who were in the timber business and they taught me a lot about how to burn, when to burn, and most importantly, when not to burn.

Dad also owned a little front end loader and we started burning several areas a year around the farm on a rotating basis. We've been burning ever since and it has done wonders for the farm. It's amazing what the burning has done for new growth and new browse for both the deer and the turkeys. It has really opened up a lot of thick areas for new browse. Some of my friends think I'm a pyromaniac, but burning has become an important part of our management program. A lot of times after we burn an area the woods will still be smoking and the turkeys will be right in there scratching away.

As an added benefit, burning has helped us with tick control. Our part of Georgia is almost legendary for those little seed ticks that can drive you crazy, and we learned a long time ago that we see far fewer ticks in the woods after an area is burned.

Basically, burning kills most of the thick brush near the ground on certain plants and promotes new tender growth on those plants that the deer love. The potash from burning goes back into the soil as a free form of fertilizer and gives the new growth a real boost. We usually start burning in late December and continue right up until late March as long as it stays cool. The Georgia Forestry Commission has been a great resource and a big help to us. To burn correctly in Georgia, you have to obtain a permit from the Forestry Commission and notify the Sheriff's Department on when you plan to burn. When we burn on the Banks Farm, we usually have a minimum of three or four people helping out and we stay in constant communication by radio.

On any given day when we plan to burn, we burn anywhere from 20 acres to 200 acres, depending on what our objectives are and how long we anticipate the job will take. We like to burn when the humidity level is low. Wind direction is critical. We like to burn when the wind is steady and the temperature is in the 50 to 60 degree range. The hotter the outside temperature is, the hotter your fire will burn. We try to use the road system on our farm to block off the areas we plan to burn so that we can control the fire better. If there is not a convenient road in a certain area, we'll cut a fire break. Whenever we burn we're extremely safety conscious.

If you would like to do some burning on your deer lease but don't have the experience to do it yourself, your state forestry commission will come out and help you burn for a fee. I've always found our forestry commission professionals in Georgia to be very knowledgeable and very good at what they do. They'll cut fire breaks around the property and show you everything you need to do to burn safely and successfully. They'll also help you plant trees if you have the need. Most states have a forestry commission and a statewide forestry association. Both organizations are excellent resources for landowners and hunting clubs. They can help your whitetail management program in countless ways.

The advantages to a well-planned burning program are numerous. Burning creates great quality browse for the deer and turkeys, and it opens up large areas to new growth.

Taking Management Bucks – Should You or Shouldn't You?

Although everyone seems to have an opinion on this subject, it can really open up a can of worms. For years, the debate has raged about shooting "inferior" spikes. Deer managers used to recommend shooting spikes on sight. Nowadays, the thinking has changed. Depending on the level of nutrition, many biologists now say that a first year buck with spikes is capable of growing a decent set of antlers in the future. Today, very few hunting clubs are actively trying to cull their spikes bucks like they used to do. But what about so-called "management bucks?" And when does a hunting club reach the level when it becomes necessary to start shooting management bucks? These are two tough questions.

Obviously, long before you even think about culling any bucks from your herd, the first order of business is to shoot enough does. I can't emphasize that enough. But when do you know when you've reached that point? This is also a question that is sometimes hard to answer. You can do a deer census on your property. That might give you an answer. There are several recommended methods for doing a census. But some are very time consuming and some are very labor intensive. With a tract as large as ours, it's tough to do a complete and accurate deer census.

In 2001, we did a partial deer census on the lower part of our farm. We consider this area to have the highest deer population on the farm. My next-door neighbor, Charles Hendrix, also did a partial census on his family's farm. Both censuses were done using trail cameras. Dr. Grant Woods, a well-known whitetail biologist who did some of his graduate work at the University of Georgia and who now lives in South Carolina, told us exactly how to conduct the census. The results showed us that we had 28 deer per square mile on that particular part of the farm. This was pretty much in keeping with the deer population estimates in this part of the state. The census also showed our buck to doe ratio to be one buck for every .88 does. That's a very good buck ratio. We were a little worried that we might have made some mistakes while doing the census, but after discussing it with Dr. Woods, he assured us that we had done everything the right way.

As mentioned in the beginning of the book, back in the early '90s when our deer herd was extremely overpopulated, we started shooting approximately 5 does for every 100 acres. Today we maintain a yearly quota of shooting about three does for every 100 acres. We try to gauge our deer population in a number of different ways. We count the number of does that we see coming out into our food plots, we examine the browse out in the woods for signs of over-grazing, and we get professional advice from experts like Dr. Grant Woods about the estimated numbers of deer per square mile in our area. This gives us a rough idea of how many does we need to shoot each year. Some years we might go a little lighter on the does, and other years we might go a little heavier.

In any event, it took us 11 years to get to the point where we thought we were shooting enough does each year. As far as the concept of shooting management bucks went, we weren't paying too much attention to that idea back in the early days. We were too busy with other concerns. As time went by, however, and as our buck population started showing increases in body and antler sizes, we started noticing some bucks that made us wonder. For instance, we might see a very heavy buck in the 185- to 200-pound range that we estimated to be at least 3 1/2 years old, and he might only have a six- or seven-point rack. We knew a buck like this would probably never grow a true trophy rack, and yet we didn't want to make the mistake of killing him just for the sake of popping a cap. So what do you do in a case like that?

This is a very sensitive issue. Oftentimes, a hunting club starts talking about culling "inferior" bucks. Pretty soon, some of the hunters in the club are shooting young bucks that might have a broken main beam (due to some type of accident), or, bucks that might have a slightly deformed antler (also due to some type of injury). Chances are, the buck that had a broken main beam will be fine the next year. He might even be trophy material. Even the buck with a slightly deformed rack might always have something odd about one or both of his antlers, but he might grow to be a non-typical monster. Yet he might never get the chance because someone decides to cull him from the herd at any early age. I've seen this happen over and over again with so many other hunting clubs. For many hunting clubs, the idea of culling problem bucks is just a good excuse to kill something.

We didn't want that to happen on the Banks Farm and we were ultra-conservative about shooting cull bucks. First of all, you have to reach the point where your deer population is down to a manageable level, and you can only do that through shooting does. Then and only then should you consider shooting a few management bucks along with the does that have to be taken. We feel like we've just reached that point during the last couple of years.

Video cameras have been a tremendous aid in helping us determine if a buck should be culled or not. During the last few years, several of our hunters have seen bucks in the woods that probably needed to be taken out of the herd, but they weren't 100 percent sure so they didn't pull the trigger. By getting good video footage of these bucks instead, and showing the footage to everyone else, we were able to evaluate each buck on the TV screen and collectively decide whether or not that buck needed to go. In some cases, we decided to take a particular buck out of the herd the next time someone saw him. In other cases, we decided to give a certain buck the benefit of the doubt. The video footage was also an invaluable aid because it showed every one of our hunters which bucks needed to go. Identifying these bucks in the woods was much easier for our hunters since they had already seen the deer on the TV screen.

In recent years, we have taken a few management

Other Management Considerations

This is the kind of buck we consider to be a cull buck on the Banks Farm. The buck was a main-framed 4x4 with very short tines. Taken by former club member Bruce Coleson, the buck was 4 1/2 years old and weighed-in at 210 pounds (live weight). He only grossed 124 points. A buck like this probably will never be a bona fide trophy.

bucks off the farm, but only a few – no more than three or four at the most. Most of these bucks were at least 3 1/2 years old, and by that time you can pretty well tell if a buck is ever going to do anything with his rack. That brings up a good point about spikes. Instead of shooting a 1 1/2 year old spike because he appears to be inferior, we'd much rather give him a few years to see what he'll do. A buck fawn born late in the season might grow a set of spikes when he's 1 1/2 because he got off to such a slow start. Our attitude is, let's give him some time to see what he'll do with a high-protein diet. If he turns out to be a genetic six-pointer with five-inch bases, we may decide to exit him from the herd when he's 3 1/2 years old. If we do decide to take him out of the herd, we might let one of our younger hunters shoot him – someone who has never killed a buck before. This philosophy has worked very well for us.

Predator and Hog Control

Predators have never been much of a problem on the Banks Farm, but in some parts of the South, coyotes, bobcats and even wild hogs can be very detrimental to your deer management program. For instance, in some parts of Alabama, I know that coyotes and bobcats are a much larger problem than they are in central Georgia. We've seen some evidence on our farm that coyotes do occasionally take young fawns, and we know that they also get some of our turkeys. In fact, they're much harder on our turkeys than they are on the deer. I've actually

My dad Lamar Banks proudly shows off the first turkey ever taken off the Banks Farm. The big gobbler was taken during Morgan County's first turkey season in 1980. Since those days, we've come a long way in managing our farm for both deer and turkeys.

had coyotes come in several times while I was calling turkeys. In each instance, I was able to do away with the coyote that came in. As a general rule, we shoot coyotes on the farm whenever we get the chance, but we've never had a big enough problem to warrant trying to trap them or kill them in any other way.

We've seen very little evidence of bobcats on the farm as well. They, too, will occasionally get very young fawns, and they can really play havoc with the turkeys. Fortunately, we have a very good turkey population on the farm and they seem to have very few problems with predators. One reason for this could be our controlled burning policy. Since we burn so many of our wooded acres, these areas stay opened up throughout the year. This gives the turkeys the advantage when a predator like a bobcat is trying to stalk them.

Hogs can be a real nuisance and they can literally destroy a food plot. Although we've never seen any on our farm, wild hogs have been seen on nearby properties. I've been involved with some land in other parts of Morgan County where we've had to trap hogs because they were doing so much damage to fields and food plots. Even out in the woods, they're just like vacuum cleaners, sucking up every acorn they can find. They're also very prolific, often having two or three litters a year. This many hogs running around loose can root up a lot of your open areas.

One of the best ways to get rid of wild hogs is to trap them. Billy Young, one of our hunters, actually traps hogs for some of the local landowners in Morgan County. Billy is an excellent all-around outdoorsman, and he's very good at what he does. Because of the problems that some of my neighbors were having with hogs several years ago we bought a hog trap. During one 10-day period, we actually trapped 18 hogs off one piece of property. Once you trap them out, they seem to disappear for a while. Then they'll start showing up again several months later.

Landowners in South Georgia and other parts of the south seem to have a much bigger problem with coyotes and hogs than we do in central Georgia. No matter where you hunt, if you have a problem with predators or hogs, you've got to stay on top of that problem before it gets out of hand. Many landowners will ask you to shoot hogs on sight, others invite hunters with dogs or traps to come in and get rid of as many as possible.

Don't Be Afraid To Get Professional Help If You Need It

I've mentioned Dr. Grant Woods' name several times in this and previous chapters of the book. Dr. Woods is a highly respected whitetail biologist and consultant living in South Carolina. During the summer of 2001, Charles Hendrix and I pooled our resources and asked Dr. Woods to come down and take a close look at both of our management operations. For a number of years now, Charles has been doing a lot of the same things on his farm that we've been doing on the Banks Farm. Our trophy management programs more or less complement each other, and we both wanted to find out from a real professional like Dr. Woods what we could do to fine-tune our management programs to an even higher level. We asked Dr. Woods to make any recommendations he could think of that would help improve our management strategies. Among other things, Dr. Woods is a recognized authority in the South on planting high-protein food plots for deer.

I had met Dr. Woods at several Quality Deer Management Association meetings and I had read a lot of his articles. In my opinion, he tells it like it is. Since I'm just a simple-minded deer hunter, he talks in a language

I can understand. I have a lot of respect for his opinion, and Charles and I were both anxious to see what he had to say.

Dr. Woods was impressed with what he saw on both our farms. He indicated that we were doing an excellent job with our food plots and our overall management programs. Because our buck to doe ratio appeared to be so favorable, he did recommend that we might start taking a few "management" bucks. As mentioned earlier in this chapter, this was something we had already been thinking about and his advice was well taken. He also recommended some things we could do to rotate some of our food plots so that they'd produce maximum tonnage for the deer.

My little brother Lee Banks shot this 110-class eight-pointer in a food plot in the mid-'90s. It was his first "nice" buck. Although our trophy standards are much higher, we sometimes allow our younger hunters to take smaller bucks – particularly if it's their first buck. Part of our philosophy on the Banks Farm is to groom young hunters for the future.

Whitetails are not true grazers like cattle. Instead of consuming large quantities of food that may not be very nutritious, deer instinctively choose more digestible plants while they are eating if they can. (Obviously, if they are starving, they'll eat almost anything.) Deer choose the most digestible part of the plant, usually the newest, most tender growth. This part of the plant also contains the most protein. (This is true with natural foods out in the woods as well as with food plots.)

A high-protein diet is so important to every deer in the herd because researchers have found that antler growth requires the same amount of minerals that does require for nursing their fawns. In fact, a growing fawn requires more protein than a mature buck needs to grow a set of trophy antlers. If a doe is not getting enough nourishment, she will not produce as much milk, and her fawns will grow at a much slower rate.

Dr. Woods explained to us that to deer, plants function as nutrient transfer agents. If the nutrients are not in the soil, the plants cannot transfer them to the deer. Antlers are approximately 50 percent protein, 45 percent calcium and phosphorous at a 2:1 ratio. If there is a shortage of nutrients in the soil, a buck's antlers will not develop to their full potential.

Having Dr. Woods evaluate our farms was certainly worth every dime we spent on his services. Ironically, while he was visiting our farm, he asked me what my personal management objectives were.

"We'd like to be able to grow a 170-class buck on the Banks Farm," I answered.

"Judging by what I've seen, it's certainly possible," he commented.

Several months later on opening day of gun season, we took a buck off the Banks Farm that netted 172 3/8 typical B&C points. When I called Dr. Woods and told him the news, he was elated.

This is what trophy hunting is all about – waiting for the right opportunity to see Mr. Big! This is a 144 B&C point power buck!

"What are your goals now?" he asked me.

"To grow a 200-class buck," I answered. "Or how about a new world record!"

All joking aside, if you think you need some good advice from a true professional like Dr. Grant Woods, don't hesitate to call someone. There are many highly qualified whitetail consultants living all across the Southeast, and they are usually very reasonable with their fees. While we're on the subject, if you're interested in deer management, I would heartily recommend joining the Quality Deer Management Association if you are not already a member. I recently became a life member. The QDMA is a great organization, and you can get a lot of good advice on quality deer management through the magazine and other publications they put out.

Maintaining an Established Program

After 12 years of intensive management, we feel like our planting and supplemental feeding program is in great shape. We feel like we have all the food plots we need at this point. We also feel like we have an adequate number of feeders for our supplemental program, and our mineral licks have been in good shape for many years. Although our program seems to be in great shape right now, we're always looking for new ways to do things and new ways to make it even better.

Over the past 12 years, each one of the guys who are members of our club on the Banks Farm has invested untold numbers of hours in sweat equity. During the early years of our program, we had old fields to clear, food plots to put in, feeders to build, and all sorts of other labor-intensive work to do on the farm. Even though we have an established program now, a lot of yearly maintenance is necessary to keep our farm in top shape. All of our members do their share in various ways. Some help out with the food plots. Others are in charge of building and maintaining some of our permanent box stands. Billy Young, whose name I've mentioned several times before, is in charge of maintaining our trail camera program which I'll talk about more in Chapter 9.

The point is, even though our management program seems to run like a well-oiled machine, there is always a job that needs to be done around the Banks Farm. And each of our hunters contributes some special talent to get that job done during the year. This is what makes our program so unique. The money that each hunter spends on his annual dues is very important to the program as well. By far, however, the contribution that each man makes through his individual talent during the year – his sweat equity, if you will – is the thing that really sets the Banks Farm apart. I'll talk more about that in Chapter 10.

If we can continue to keep our food plots in good shape by spraying them for weeds and insects and making sure they have adequate lime and fertilizer, if we can keep shooting does every year like we've been doing, if we can keep filling in those protein gaps during the year, and if we can just stay on track with our existing program, we should keep growing and harvesting some real power bucks for years to come. That's really what it's all about!

Summary — Chapter 8

- Maintain food plots - summer and winter. Spray for weeds, insects. Check for lime and fertilizer.
- Maintain supplemental feeders. Make sure feed pellets stay dry and clean.
- Maintain mineral licks.
- Plant for turkeys, quail.
- Maintain farm roads.
- Bush-hog old fields as necessary.
- Make sure you are filling protein gaps from January through August.
- Protect younger bucks.
- Shoot enough does to maintain over all program.
- Take steps to control predators, if necessary.

Would you believe this is a Boone and Crockett buck that scores 172 3/8 typical B&C points? We didn't either, until the opening day of the 2001 rifle season. This is the Horn Donkey in all his glory. His antlers grossed 197 B&C points including abnormal sticker points. (photo courtesy of Charles Hendrix)

Chapter 9

Using Trail Cameras and Video Cameras

Almost every serious deer hunter owns a video camera these days. Every season, hunters get some incredible video footage of big bucks in the woods. Trail cameras have also become very popular with hunters and hunting clubs in recent years. Like many other clubs, we started using trail cameras just to see what might turn up. We never dreamed they would become such an important part of our management program. In fact, I consider the use of trail cameras and video cameras so important that I wanted to have a separate chapter in the book devoted to this subject alone.

Back in 1996, Billy Young, one of our club members who I've mentioned several times, started putting out Cam Trakker trail cameras as a way to monitor some of our deer and get night photos of bucks at some of our feeders. We quickly realized what a great asset these trail cameras could be. Since that time, the cameras have given us the opportunity to get pictures of certain bucks that we didn't often see during the day. In some cases, we were actually able to partially pattern certain bucks from the location of the cameras around the farm. (The Horn

It takes a real knack to set up a trail camera in the right spot and at the right angle. Banks Farm club member Billy Young is a master at putting our trail cameras in the right places. Billy got this outstanding shot of three longbeards strutting.

Donkey was one of those bucks!) We also got to see pictures of certain "ghost" bucks that were living on the farm. You might get one picture only of a certain buck, and then you'd never see him again after that.

Trail cameras have been a great teaching tool for all the guys who hunt on the Banks Farm. We often study the various photos taken in order to determine things like the ages and weights of certain bucks, and of course, the size of a buck's antlers.

Billy Young runs our trail camera program today. He's not only a great hunter, but he's a real master at setting the cameras up in the right places. It takes considerable practice to set the cameras up at the right angle and at the correct distance from the subject. Most of the cameras on the Banks Farm are set up over feeders because that's where the deer are going to be. Occasionally, we'll put a camera out near a well-used trail. We currently keep

about six to eight different trail cameras going from late summer to late winter. On the Innisfail Farm next door, Charles Hendrix also keeps several trail cameras out on his property during the same time period. Since our farms share a common boundary, we often get pictures of the same bucks.

As mentioned in Chapter 1, the first time we ever saw the buck that became known as the Horn Donkey was in a photo taken with a trail camera at a feeder. During the later summer and early fall of 2001, we actually got seven or eight different photos of the Horn Donkey with different cameras set up at different places on our farm. Charles Hendrix also got several different pictures of the Horn Donkey with his cameras. From these photos we began to see a definite pattern as to the trails the Horn Donkey was using on a regular basis. This knowledge played a key role on opening day of rifle season in 2001 in helping me decide where to hunt that day. As luck would have, I made the right decision. The Horn Donkey made his appearance near the stand where I was hunting

Our next-door neighbor Charles Hendrix got these three exceptional night shots on the Banks Farm with several different trail cameras. (Top left) a shot of the Horn Donkey taken on September 30, 2001, about a month before he was killed; (top right) two nice bucks fighting; (bottom right) a camera-shy 10-point buck crouching.

late in the afternoon. (See Chapter 11 for the complete story of the hunt for the Horn Donkey.)

Trail cameras cost between $275 and $400 each, but they are well worth the investment. When you consider the fact that your trail cameras can produce excellent photos of big bucks like the Horn Donkey, the cost of the film and processing is also a very worthwhile investment. Many trophy clubs complain that although they pass up numerous 1 1/2 year old and 2 1/2 year old bucks each year, they never see any older age-class bucks in the daytime during hunting season. I believe there are definite reasons for this problem, but the best way I know of to find out what kind of bucks you have on your property is to put out several cameras. You'll be amazed at what you might see. Chances are, you'll get pictures of bucks you never knew you had on your property.

The Video Hunter

My cousin Danny Keever, who I've also mentioned several times, started operating a video camera on the farm back in 1993. Since that time, Danny has gotten excellent footage of numerous bucks in the 130- to 140-class range. I believe Danny gets as much enjoyment out of shooting video footage of big bucks as he does shooting his rifle. Over the years, Danny has videoed far more big bucks than he has taken with a rifle. When people ask him why, he just smiles and says, "One of these days, when I see the right buck in the woods, I won't be holding my video camera!"

Danny's enthusiasm for shooting excellent video footage of big bucks sort of grew on everyone else in the club. Various hunters would come back in after a day in the woods and say, " Man, I wish I had had a video camera with me today." Then they would tell everybody about some big buck they had seen in the woods.

Eventually, most of our guys began acquiring video cameras. Today, our hunters seldom go to the woods without a video camera in their possession. Most of the cameras we use are the small "Hi-8" digital cameras because they are very portable and they're not nearly as bulky as the larger cameras. These smaller cameras suit our purposes just fine.

In 1999, Carl Wilson videoed a beautiful 10-pointer during the rut that would have easily grossed in the 140s. Unfortunately, he didn't have his bow with him at the time. The buck stepped within 17 yards, but he decided not to shoot it with his rifle. If he'd had his bow, he no doubt would have ended up with a fine Pope and Young qualifier. Three or four days later, Danny Keever saw the same buck several hundred yards down the creek from where Carl had seen it. Danny had already seen Carl's video footage and he recognized the buck immediately. Danny shot some good video footage of his own. In fact, he saw the buck several more times that season and he got good video footage each time.

Although this particular buck was a definite shooter under our trophy guidelines, no one in our club ever got the opportunity to shoot it. To our knowledge, we never saw it again. This is just one example of how video footage can help you follow a certain buck during the season.

In 1994, the year we had so much rain, Danny Keever got great video footage of four big bucks out in the same food plot together. Each one of these bucks scored between 140 and 160 B&C points. The 160-point buck was later killed. Another of the bucks, scoring 150 7/8 B&C points, was later found dead.

If someone sees a buck on our farm and they decide to pass it up, they always try to get some good video footage of that buck to show everyone else back at the cabin. Even if someone sees a buck they intend to shoot, they try to get footage of it before they pull the trigger. This really adds to the excitement of the hunt. Our standards for shooting bucks are high on the Banks Farm. A lot of our hunters consistently pass up 130- to 140-class bucks. Being able to study older age-class trophy bucks of this size on video is an added bonus for the entire club.

Like the still photos taken by the trail cameras, the video footage we've gotten over the years has been an awesome teaching and coaching tool for all of our club members. Over the years, we've gotten video footage of all kinds of critters – bobcats, fox squirrels, turkeys, and raccoons. Whenever someone shoots some good video footage of a certain buck, we all get together around the TV and study the footage. Just like with photos taken by our trail cameras, we try to determine things like: *Is he a shooter? How old is he? How much does he weigh?* And, of course: *What will his antlers score? How long are his tines? What's his inside spread?*

We've all learned a lot from the various videos taken by different hunters. Sometimes we find that our educated guesses about antler sizes and weights are way off the mark. Other times we're right on the money. If a question comes up about whether or not a certain buck should be culled as a "management" buck, we have had situations where video footage of that buck was taken and studied by several club members before any final decision was made in regard to shooting that buck.

Watching videos has also been a great learning tool for our youngest hunters. Some of our club members have children who are just learning how to hunt. Video footage shot by our hunters is often used to teach these children what constitutes a trophy buck on the Banks Farm and how to judge antlers and body sizes. Video footage also teaches young hunters how to identify button bucks. Since many of our youngsters begin their deer

Prints like this made from video footage can tell you a lot about the bucks in your herd.

Here's another Danny Keever print made from video. Two nice bucks!

hunting careers by shooting does, this knowledge helps prevent mistakes from being made when doe-shooting time arrives.

Quite often, we've gotten video footage of the same bucks year after year on our farm. In fact, we've gotten good video footage of several 140-class bucks two or three times during the course of a single season. Like I've said before, this indicates to me that our mature bucks really aren't traveling too far off our property, and that's good news. In the past, I've read stories about trophy bucks that disappeared during hunting season and were later seen many miles away from their core areas – maybe 12 to 15 miles away. These bucks reportedly never returned to their core areas until hunting season was long gone. I'm sure this happens, and I'm sure there are reasons why this happens. On our place, however, many of our bigger bucks seem to stay a lot closer to home.

Deer will always travel to where the food sources are. When the acorns begin to disappear in the hardwoods, the deer will move on to the next best food source. That may be a food plot two miles away. Between the quality food plots and the cover we have on our property, our deer never have to go far to get all the good food they can eat. I strongly believe that is one of the primary reasons why many of our bucks don't seem to travel too far and why we consistently see them over and over again in the same areas.

Trail cameras and videos cameras have become an important part of our deer hunting operation. I strongly recommend using one or both of these tools to help you keep track of some of the different bucks on your property.

Summary — Chapter 9

- Trail Cameras set up over feeders in the woods at night often produce pictures of older age-class bucks you never knew you had on your property.
- The use of video cameras by club members while they are hunting can be a huge asset to any serious trophy management program. Videos are a great teaching tool. They can help you learn many important things such as how to judge trophy antlers, how to identify button bucks, or how to identify potential "management" bucks.

Club member Pat McDevitt and I go back a long way. We played football together at Georgia Southern University. Pat is a real asset to our club. Here he shows off his 145 B&C point power buck taken in 2001. It's so important to make sure all your club members are compatible.

Chapter 10

Goals for Your Hunting Club

Choose Your Members Carefully

I've mentioned several times in previous chapters the importance of having a compatible group of hunters in your club who share like-minded goals. We've been extremely fortunate on the Banks Farm to have a close-knit group of hunters who don't mind hard work. Each man in our club has his strong points and talents. Each man uses these talents in various ways to achieve the annual and long-term goals that we set. Each man also goes by the rules. More importantly, each one of our hunters has a passion for what we are trying to achieve, and that passion shows in his everyday attitude.

Like almost every hunting club in existence, we've had members in the past who didn't work out for one reason or another. Whenever this has happened in our club we've taken quick steps to remedy the problem. For instance, say that all the members of your club are serious trophy hunters and you've set some high standards for shooting bucks. Say you have one member who can not seem to refrain from pulling the trigger whenever he sees a smaller buck. I'm not talking about making an honest mistake. We all make genuine mistakes occasionally. I'm talking about the hunter who looks for any excuse to pull the trigger. You can't have a hunter like this in your club and expect to achieve your goals.

One bad apple can really upset the apple cart, and it's very important to recognize this fact before things go too far. If you have a hunter in your club who can not abide by your standards, that hunter needs to join another club. It takes a lot of restraint to be a serious trophy hunter.

It took a lot of hard work to get to the point where our club is today, and one of the main reasons we've been able to do it is because we have such a compatible group of guys. Everyone has a niche that they fill and everyone is on the same page. Some of the guys help with the planting and the maintenance of the food plots. Some help with the feeders and putting out minerals. Others help with the maintenance of the cabin.

Communication among the various members of a club is so important. On the Banks Farm, we talk among ourselves a lot. If someone has a good idea about something, we listen to that idea and we discuss it. If we have a disagreement about something, we hash that over, too. In the end we try to let common sense prevail.

Choose the members of your club carefully. Make sure everyone is compatible and shares like-minded goals. Make sure every member contributes more than just his annual club dues each year. By contributing his fair-share of sweat equity to important projects throughout the year – projects like planting food plots, mowing and bush-hogging, maintaining farm roads, clearing old fields, building and maintaining permanent stands, and maintaining the hunting cabin to name only a few – a club member will build a real sense of ownership in that club and he'll build a much closer bond with his fellow members. More importantly, when he shoots that buck of a lifetime, it'll mean more to him.

If you do happen to get a club member who does not want to do his part or who refuses to go along with the basic club philosophy, encourage him to find another club to join. Your overall program is too important, and your fellow club members have too much at stake to let one bad apple ruin things for you.

The Banks Farm "Guys"

Since everything we've done on the Banks Farm has been a team effort, I wanted to include a section about the guys I hunt with. I've already mentioned several of the guys numerous times in other chapters. In sports, people love to remember famous teams or individuals that accomplished some incredible feat – like the U.S. Hockey team in the 1980 Olympics that came charging out of nowhere and beat the Russian team. If you saw it on TV that was a moment in sports history that you'll never forget.

I hope this doesn't sound too sentimental, but to me, the team of hunters we formed back in the early '90s sticks in my mind in a very similar way. Although several of the original members are no longer hunting with us, that team of hunters who joined our club during those early years has materialized and grown into something very special over the years. It's been fun building this team and growing with it, and I wouldn't trade the memories we've created for anything in the world. Right now we have active 12 members in our club including me. That's a good number for the size tract we hunt. We seldom kill more than four trophy bucks off the farm in any one year. Several of the guys who hunt with us have never killed a buck off the farm. They're still waiting for their chance.

I met Carl Wilson at Gaithers United Methodist Church over in Newton County back in the late '80s. I always like to label Carl as the first real Christian friend I ever had. To say that Carl is mechanically inclined would be an understatement. When it comes to running dozers or tractors he's a pretty talented cat. In fact, Carl and I do most of the tractor work on the farm. He's also a very good hunter and an all-around outdoorsman. In 1990, Carl came over and started helping us shoot does. When I first met him, his little girl Carlynn was just a baby. Now she's a teenager and a beautiful young girl and she hunts and fishes with her father all the time. Carlynn's already killed several deer on the farm. During the spring gobbler season in 2002, she killed her first gobbler – a fine longbeard.

One of my fondest memories of Carl goes back to 1990 when he offered to come over and help us shoot some does. He showed up with a complete arsenal of weapons. He had a Browning .300 Magnum A-bolt rifle; a muzzleloader; a .22 Magnum rifle in case he saw some squirrels; and he had one of those huge Ruger Blackhawk .44 Magnum pistols with a scope on it.

When I first saw him with all those guns I said to myself, "Either this guy is totally crazy, or else he's a good hunter." He came back with his limit of does that day and he proved his worth as a hunter. We've been close friends ever since. Carl is an invaluable asset to our club. In fact, all of our guys are. That's why we are so fortunate. We have a great set-up. We're just a group of normal guys who have some common goals and we work well together in achieving those goals. We don't have any game hogs in our club. Whenever someone kills a nice buck, everyone else is genuinely happy for them. I've watched Carl's kids grow up with mine and it's been fun.

Another long-time club member is Pat McDevitt. We met while we were playing football together down at Georgia Southern University in 1983. One weekend while we were still in school, Pat happened to tell me he was going deer hunting on some family property in Morgan County. I didn't even know he was a deer hunter until then. To show you what a small world it is, it turned out that Pat's uncle owned a farm located only a few miles down the road from our farm. While we were going to school, I never had anywhere to hunt around Statesboro. I'd always start moaning when deer season was about to start because I couldn't stand the thought of not being in the woods. Pat had two or three good places to hunt and he started taking me with him. We've been hunting together ever since.

One Friday afternoon in October during football season, we were having a team meeting with Coach Erk Russell. Pat and I were both linebackers. Pat was a middle linebacker and I was outside linebacker. We were going over scouting reports and other pre-game information. Coach Russell knew that Pat liked to deer hunt a lot. In fact, Pat was every bit as much of a whitetail fanatic as I was.

Coach Russell looked over at Pat and said, "Hey, McDevitt!" Pat answered, "Sir." Coach Russell said, "Do you know what tomorrow is?" Pat said, "Yes, sir. It's Game Day!" Coach Russell repeated, "No, do you know what tomorrow is?" Pat said, "No sir. I don't know what you mean." Coach Russell said, "Tomorrow is opening day of deer season. (It was also Game Day.) Don't you damn go!"

I'll never forget that. Pat wasn't about to miss out on opening day, though. I can't swear to it, but I'm pretty sure he snuck out in the woods for a little while that morning despite Coach Russell's warning.

A boy's first buck is something he'll never forget. A smiling 14-year-old Travis Keever, son of club member Michael Keever, shot this long-tined 10-pointer on in December 2001. The buck netted 134 B&C points. By now, you probably realize how much emphasis we place on family participation around the Banks Farm.

I've mentioned several times that my family has always been a very close-knit family. When I was growing up, three of my cousins – Bert Clotfelter and Danny and Michael Keever – hunted with us on a farm we had over in Newton County. In 1993 Danny joined our club on the Banks Farm. The next year in 1994, Michael came on board as well. Danny and Michael trim houses for a living and they can build anything. They build most of our feeders, and all our permanent "Texas" tower stands. They also take care of any necessary maintenance around the cabin. In addition to building and maintaining our feeders, Danny helps with putting out feed and minerals. Since it's been so dry in recent years, Danny's had his work cut out for him just keeping the feeders filled up.

Like a lot of clubs, we had an old house on our property that we used as a cabin for years. Several years ago, we totally renovated the cabin and made a really nice house out of it. We added on a large family room and a large back deck. Danny and Michael did most of the carpentry work. In addition to his building skills, Michael is the chief chef for the Banks Farm hunters. He loves to cook, and he can prepare anything from fresh fish caught out of our lake to grilled steaks that melt in your mouth.

My cousin Bert Clotfelter also joined our club in the mid-'90s. As a boy, I probably spent more time with Bert than I did with Danny or Michael. Bert was always hunting something – crows or some other kind of critter. We spent a lot of time together in the woods. Back in the middle '90s, Bert earned his stripes with the club by passing up some really good bucks. He passed up several nice 10-pointers that easily would have scored in the mid-130s. Being a good role model is part of being a good trophy hunter. Bert has proved his worth to the club many times over.

Bert does a lot of reloading and he's helped us all out with his excellent reloads. On the Banks Farm, we believe in good shooting. We're big on making sure our rifles shoot a 3/4 inch group at 100 yards. In fact, we like to see a three-shot group where all the holes in the target are touching each other. You might only get one chance at the buck of a lifetime, and you'd better make sure your rifle is right on the money. In the early days of the club, Bert was very instrumental in making sure we all had bullets that would drive nails. Pat McDevitt also does a lot of reloading. Between the two of them, they keep our rifles in top shooting condition.

Joe Chandler and his wife Karla joined our club about two years ago. I met Joe through a mutual friend several years ago. We immediately discovered that we had a lot in common through our interest in hunting. Joe owns a farm in Upson County. He also owns several hundred acres in Morgan County. He's big into trophy management. He's also very knowledgeable. We're always comparing notes about food plots or other related topics.

Shortly after we met, Joe invited me down to his farm in Upson County and showed me all the things he was doing with food plots and supplemental feeding. He helps out around the farm by keeping our feeders filled up during certain times of the year. He and Karla are both serious trophy hunters and great assets to the club.

Jarrod Brannen is another very close friend. I met Jarrod at school down in Statesboro. Jarrod didn't play football with us, but he's very athletically inclined. He stands about 6-foot, 3-inches tall and he likes to work out with weights. We met in the weight room one day while we were both working out. I found out that he was interested in deer hunting, and we got to be good friends.

Like Pat McDevitt, Jarrod joined our club back in the early days when we were just getting started. In '93, Jarrod shot a beautiful 11-pointer (a 5x5 with one sticker) that grossed in the mid-150s. Jarrod's buck had good tine length and a spread of about 17 inches inside. It weighed 220 pounds (live weight). As of that date, it was the heaviest buck killed off the farm. (In '96, I killed a buck that weighed 224 live weight.).

Jarrod's buck was also the largest buck killed off the farm as of that date. It was an older age-class buck (probably 4 1/2 years old), and it obviously had some great genetics. Unfortunately, we had a drought in '93 and everything was extremely dry that year. The only thing Jarrod's buck lacked was exceptional antler mass. We had not really perfected our high-protein feeding program at that time, and I often wonder what Jarrod's buck would have scored if it had been getting all the protein it needed.

Jarrod's a good deer hunter. In 1996, he killed a huge buck in South Georgia (Dooly County) that netted 160 6/8 after considerable deductions. One of these days, I'm sure he'll take one like that on the Banks Farm.

I've mentioned Billy Young's name several times throughout this book. He's a very talented all-around hunter. As mentioned in Chapter 9, Billy's primary job around the farm is to keep track of our trail cameras. He has a knack for putting them in the right places, and he's gotten some incredible photos of trophy bucks.

Having been a hunting guide at Burnt Pine Plantation, Billy's done it all. He's tracked wounded deer, scored antlers, and he knows a lot about food plots and whitetail management. He also knows a lot about hunting trophy whitetails. We became friends back in the early '90s when Billy was still working at Burnt Pine.

This was the first of many "power" bucks taken off the farm since we started our intensive management program in 1990. My good friend Jarrod Brannen shot him in 1993. Jarrod's buck grossed in the low 150s, and was probably 4 1/2 years old.

Billy is one of the original gang, having joined the club in '92. Back in the mid-'90s, he killed a big seven-pointer off the farm that had seven-inch bases. That deer was a cull buck if I ever saw one. He grossed 132 points, though, making him a legal trophy according to our trophy standards.

Pat Hayes, an old high school buddy of mine, joined the club in '99. We grew up together and I spent a lot of time with him and his family during our high school years. Pat's family owns some land in Newton County which they've had under trophy management for a number of years. Several 140- to 160-class bucks have been taken on that property, so Pat knows what a trophy buck is. During the short time he's been a member of our club, he's seen and passed up several very good bucks. In 2001 he saw a very big buck but could not get a good enough look at it to determine how big it was. So he let it go.

Pat's family is in the heating and air conditioning business. When we remodeled our old farm house, Pat helped us with installing the heating and air systems. He also helped us build our deer cooler. Pat's company has a prefab shop, and they can make anything. He's made a number of first-class, lock-on ladder stands for our use on the farm. Pat's another good all-around hunter who we are grateful to have in our club.

A newcomer to the club is another cousin of mine, Darrell Flanigan. He joined the club in 2001. Darrell is one of our youngest hunters, so we give him a hard time about everything – especially passing up trophy bucks. Since he's a newcomer, we're always on his case about videoing nice bucks instead of shooting them with a rifle. I think we've made him a little paranoid about *not* pulling the trigger. So far, he hasn't shot anything with his gun, but he will one of these days. He's a great young man. He's also in the building business like my other cousins. Last year, Darrell built several new feeders for the farm. Darrell is one of those people who always joins in and helps out with any job. He's a pleasure to have around.

While I'm on the subject, I also want to mention two other people who have been extremely important to our club. The first is Mr. Neal Vason. My family has been leasing a piece of property from "Mr. Neal" for about 20 years now. The property is adjacent to our farm and having it leased has really helped us with our trophy management program because it more or less gives us one big block of land to work with.

Mr. Neal has always treated me like one of his own. He's allowed us to put in some quality food plots, keep the roads cleaned out and do other things to his property just like it belonged to us. I really appreciate the confidence he's placed in us and I value the relationship I have with him. If you lease your hunting land like many hunters do, having a good relationship with the landowner is so important. We've been extremely blessed over the years to lease land from such an understanding and compatible landowner. It's been a privilege working with Mr. Neal.

The second person I want to mention is Charles Hendrix, our next-door neighbor. As mentioned, Charles and his family have had Innisfail Farm under an intensive trophy management program similar to ours for many years. Charles and I have worked together in many ways, and our combined efforts have really complimented each other. We're in constant communication regarding important management issues like how much feed the deer are eating on each farm, how our food plots are doing, and other related issues.

Just like with the Banks Farm, only a few trophy bucks are taken off Innisfail Farm each year. Fortunately,

Charles Hendrix photographed this big buck in 1999 with one of his trail cameras. He later shot the buck during rifle season. (See opposite photo)

we don't have to worry about our bucks walking across the property line and being shot. To have a tract as large as Innifail Farm right next to the Banks Farm has been a definite plus for our trophy program. As mentioned, we have a combined total of about 5,500 acres. Over the years, Charles and I have become very good friends. You couldn't ask for a better neighbor.

Make It a Family Affair

One of the nicest things about our place is that we've always strived to make it a family operation. Since we operate a serious trophy hunting club, no drinking of any kind takes place on our property. Children are always welcome, and we encourage our hunters to bring children, wives, or other family members any time they want to fish in our lake, or participate in other club activities.

We frequently get together and have family cookouts. We also encourage family members to help us shoot does.

We've probably had at least 30 youngsters kill their first deer or turkey on the farm. We're very proud of that record. Children often accompany their parents on both deer and turkey hunts. Sometimes various members will simply take their children out to sit in stands and video deer out in the food plots. We have children and cousins and nieces and nephews who all hunt on the farm together.

Family members are not treated as guests on the farm. They are treated as members and made welcome any time. If a club member decides to let his wife or child take his one allotted buck during the course of the sea-

Charles Hendrix has been a great friend and neighbor over the years. The way he and his family manage Innisfail Farm has greatly benefited our trophy management program on the Banks Farm. Here Charles shows off a beautiful mainframed 4x4 taken on his farm in 1999. The buck grosses over 140. Live weight was 220 pounds. The buck was believed to be 6 1/2 years old.

son, that's not a problem – as long as that member is there to supervise.

On May 15, 2002, while Duncan Dobie and I were busy working on this book, an event took place on the farm that was almost as exciting as the day I killed my Boone and Crockett buck. It was the last day of turkey season in Georgia, and my five-year old son J.L. really wanted to go turkey hunting. Because he's still quite small, I got him a single-barrel .410 shotgun and had the stock cut way down.

J.L. practiced shooting his little shotgun. To my amazement, he could shoot extremely well. But I knew there was a big difference in being able to shoot a paper target and shooting a live turkey gobbler. Since it was the last day of the season, Carl Wilson and I joined up with J.L. and my nephew, James Nail. The four of us went out together and started calling.

We immediately got an answer. Carl and I set up with the boys about 10 yards apart. J.L. was sitting in my lap as the turkey started coming in. James was with Carl. We expected the turkey to come right down the road in front of us, but big gobblers seldom do what you expect. He came around through some thick stuff to our right. Carl and James were not in a position to get a shot. Just as the turkey was about to appear, I swung around with J.L. in my lap so that J.L. could reposition the gun. For some unknown reason, the gobbler never saw us move.

As the gobbler came into view, I whispered, "Do you see the turkey, son?"

"Yes, daddy," J.L. whispered back.

"Okay son, put the gun up and aim right for his head," I whispered.

The turkey was about 26 yards away. J.L. tried to aim his little shotgun as instructed, but the stock got hung up in his shirt. When that happened, his mask came down

Here's five-year-old J.L. Banks with his first-ever gobbler taken on the last day of the 2002 spring gobbler season in Georgia. J.L. made a perfect shot – completely by himself. The shotgun had to be cut down and it was still almost too long for him!

over his eyes and he couldn't see.

The turkey was right there on us. I helped J.L. fix his mask and get the gun in position.

"Can you see him now?" I asked.

"Yes, daddy," J.L. answered.

I cocked the hammer for J.L.

"Got it on his head?" I asked.

"Yes," J.L. answered.

"Pull the trigger, son."

J.L. pulled the trigger. I fully expected that turkey to fly off or run away or jump straight up in the air. Instead, he went down like a sack of salt.

"You got him son!" I yelled. "Go get him!"

J.L.'s turkey weighed 17 pounds. He had a 10-inch

Here's a happy bunch of turkey hunters. (Left to right) Carl Wilson; my nephew James Nail; J.L. "the turkey slayer" Banks; and me, one proud – and surprised – father!

beard and one-inch spurs. I'm telling you, my heart was pounding. I never did a single thing to help J.L. shoot. He held the shotgun all by himself. He was one proud boy, and I was one excited father. Carl and James were also excited.

That's what real hunting is all about!

Establishing Minimum Gross Scores for Trophy Requirements

Back in 1990 when we first started our trophy management program, we had a two-buck limit. Both of those bucks had to be "mountable." We quickly found out that a "mountable" buck is open to a wide range of interpretation. What constitutes a mountable trophy? Does a 2 1/2 year old eight-pointer that scores 110 B&C points qualify as a mountable trophy buck? Some hunters might think so.

Even in those early days of our program, I knew that we were shooting "mountable" bucks that we should be passing up. I realized right away that we had a big problem with such a discretionary method for determining what a trophy buck was. I also recognized very early in the program that a two-buck limit was not going to work on our farm.

Fortunately Dad imposed a one-buck limit as mentioned earlier in the book, and it's been that way ever since. After the one-buck limit became a rule, we started wanting to learn more about scoring our bucks. I didn't know much about scoring antlers back in the early '90s, but I thought it was important to learn. We started out by practice-scoring some of the deer we had mounted. Our taxidermist Ricky Smith, who has a shop in Newborn, Georgia, was very instrumental in helping me and some of our club members learn the basics for measuring racks according to the Boone and Crockett scoring system. Ricky knew a lot about scoring, and he was a very good teacher. He actually held several mini scoring clinics at his shop over in Newborn for some of our hunters.

Billy Young, one of our hunters, was also very good at scoring racks. Billy had been a former hunting guide at Burnt Pine Plantation. Burnt Pine is a well-known hunting operation located about 10 miles south of our farm. While guiding there, Billy scored dozens of whitetail racks. Thanks to Billy and Ricky, everyone in our club became very proficient at scoring racks – especially on live deer out in the woods. Today, we make a point to continue that process by teaching our youngest hunters how to score racks as well.

I have never been an advocate of using spread minimums or trying to go by the number of points on one side (like the four-point rule) or using minimum tine measurements to determine what would constitute a legal buck on our farm. Instead, I always believed that we should go by total gross score. I still think this is the best way to do it for any club serious about hunting trophy bucks.

Today, things are a lot different than they were 10 or 12 years ago. Because of all the emphasis placed on the record book, practically every serious deer hunter in North America is familiar with the Boone and Crockett scoring system. *Does it make the book?* is one of the most common questions you here in deer hunting circles. If you don't know anything about scoring antlers, I strongly encourage you and the other members of your hunting clubs to learn how to score deer on the hoof so that you don't shoot the wrong bucks.

Once you are familiar with scoring mounted deer heads it gives you a whole new perspective on judging the size of antlers when you are watching live deer out in the woods. You might have a buck hanging on the wall that scores 115 points. He's got three-inch brow tines and eight-inch G-2s. After you see a similar-size buck standing out a food plot, it's easy to compare the live deer's rack with the mounted buck's rack hanging on the wall back in your den. Then you can ask yourself: *Are the tines of the live buck as long or are they longer? Does the live buck have more mass? Are his main beams about the same length? What about his inside spread?* This kind of comparison with a known score makes it easy to say, "Yes, he's a little larger," or "No, he's a little smaller."

When I was learning how to score racks, I scored every mounted deer head I could get my hands on. Then I started trying to score every live buck I saw in the woods. I made plenty of mistakes. I still do occasionally, but most of the time my estimates are fairly accurate. (One of the biggest mistakes I've made recently was under-estimating the size of the Horn Donkey's antlers from the trail camera photos we took of him. (I'm still trying to live that one down.)

After our program had been running for a few years, I decided that our trophy bucks should have a minimum gross score of 130 B&C points to make them legal trophies. I arrived at this minimum score after measuring all the bucks we had mounted from previous years. I started taking a closer look at some of these different bucks, and I realized that many of them were nice young 10-pointers that scored between 120 and 130 points. I knew that these were the very bucks we should be protecting, so we came up with the 130-point rule. Sure, we occasionally make a mistake, but this rule has worked very well for us over the years. In fact, as I've talked about several times in other chapters, most of our

Mother and son headed to the woods! This picture was taken in the early '90s when my younger brother Loy was about nine or ten years old. My mom Barbara has always been very supportive of our deer hunting obsession.

hunters consistently pass-up bucks that score well beyond the 130-point mark.

A One Buck Limit is Essential!

Having a one-buck limit is a huge point of contention with some clubs. If you or any of your hunters have to kill more than one buck every year on your club lease, you're not serious about becoming a trophy hunter. More importantly, you're not serious about protecting your younger bucks. It's that simple. In order to grow true power bucks, you've got to be willing to go several years or more without pulling the trigger. I killed an exceptional 12-point buck on our farm in 1994 that grossed 161 2/8 points. Two years later in '96, I killed a beautiful 10-pointer that grossed 150 points. I did not shoot another antlered buck on the farm until five years later in 2001.

Was it worth the sacrifice? Let me put it this way. I read somewhere that only one out of every one-million hunters ever kills a Boone and Crockett buck during any given year. In truth the percentage is probably much lower – more like one in every few hundred thousand – but still it doesn't happen every day. In 2001, I killed a Boone and Crockett buck on our farm that scored 172 3/8 typical B&C points. Would I wait another five years to do it again? You bet I would! Was it worth the sacrifice? Strange as it may sound, it really wasn't that much of a sacrifice.

I passed up dozens of nice bucks during that five-year period. With the possible exception of one or two of those bucks, I passed up all of those deer because they weren't big enough to shoot. I always knew that sooner or later the right buck – a real power buck – would come my way. He did – on opening day of the 2001 season! We knew we were growing some incredibly large bucks on

Hunting on the Banks Farm is a family affair! Always has been, always will be! Here I'm posed with my son and daughter – five-year-old J.L., and nine-year-old Kimberly – in October 2001 shortly after shooting the Horn Donkey.

our farm. For me it was just a matter of time.

That's the attitude that all of our hunters have. That's the attitude that any true trophy hunter *has* to have. On the Banks Farm, our hunters don't pay for *a* buck. They pay for the *opportunity* to hunt a real power buck! With the kind of program we have, everyone knows that the *opportunity* will present itself sooner or later!

Traffic – Keep It to a Minimum

It's worth repeating that we go to great lengths to keep the traffic on our farm down to a minimum. As far as working in the fields or putting out feed, I don't think that bothers the deer at all. They're used to seeing us driving around doing our various jobs on a daily basis.

We're out on the farm all the time doing something — planting, burning, or working on the roads. But we don't traipse all over the property on foot or go joy riding just to see what we can see. There is usually a definite purpose to everything we do.

Sometimes our hunters like to sit in stands in late summer and watch the deer out in the food plots just to see if they can spot any big bucks before hunting season starts. But these guys are very careful, and they don't go walking through the woods putting out scent all over the place. If you've ever been to a game farm you can learn a lot by watching the deer. Even though they're semi-tame, they still react to human presence. You can walk down the regular trails all day long and it doesn't bother them one bit. But the moment you step off the main trail, the deer go crazy. They know something is up. The same is true on any hunting lease. Deer accept regular activity without getting alarmed. But the minute you go sneaking around in places where they're not used to seeing human activity, they know something is up.

By keeping our traffic to a minimum, I believe we see a lot more bucks on our place during daylight hours. I believe even the older age-class bucks are more inclined to come out into the foods plots during the daytime if they don't feel threatened. There are plenty of areas on the farm where we seldom set foot, and we try very hard to never push a buck out of his core living area.

During hunting season, it's pretty quiet on the farm up until Thanksgiving. When our club members start hunting seriously in the fall, all activity in the woods is kept to a minimum. We don't ride all over the property on four-wheelers. Our hunters go to and from their stands as quietly as possible. After Thanksgiving, doe-shooting time begins and things get kind of noisy.

Communication Is So Important

Although this topic doesn't have a lot to do with goal-setting, it is extremely important. Good communication between club members is essential. In this case, I'm talking about voice communication between club members when they are physically on the club property. It seems like one of our club members is always out on the farm doing some job – even during the off-season. For this reason, we stay in constant touch by way of radio-telephones.

It doesn't matter what time of year it is. In the spring when we are planting our iron and clay peas, we're always in communication. If a tractor breaks down in a field somewhere, one call is all it takes to get help. In late winter, when we are burning, everyone who is helping out always has their radio-telephone turned on. Being able to stay in touch is an important safety tool. Even when we are on our deer stands in the fall, we usually have our radio-telephones close by.

If an unwanted guest is spotted doing something suspicious on our property, as soon as an alert call goes out several people can zero-in on that spot within a matter of minutes. If your hunting club has more than several members, chances are you'll be as busy as we are on a year-round basis. Today's cell phone technology is incredible. In the old days, we had CB radios, two-way radios, and mobile phones. We all have Nextel radios now, and they're great for our purposes. Because they're a telephone and a radio in one, I can call one of our club members as far away as South Georgia and get him in a matter of seconds. The safety issues alone are enough for us to justify having quick two-way radio communication on our farm. There are many phones on the market that have two-way capabilities. Do yourself a favor and make sure all your club members have a good way to stay in touch with each other.

Aim High – Be the Best You Can Be

I learned a lot during my football career at Georgia Southern University. My coach, Erk Russell, used to tell the team that hustle could make up for a lot of inability. We've always believed in hustle on the Banks Farm, and I think it's helped make us a winning team. I once read an article on excellence that said if better is possible, good is not enough any more. That's the kind of the attitude we've always had with our hunting club. We always strive to do things better. We've been successful through hard work, trial and error and determination. We've always been willing to try new things. If you'll try the same approach using some of the methods that have worked so well for us, you'll see results and you'll start producing some real power bucks.

Someone told me once that the definition of insanity is doing things the same way all the time and expecting different results. We've never been willing to settle for doing the same thing all the time. For instance, we're always looking for better ways to fill in those protein gaps that occur during the year. We're always trying to fine-tune our program. We used to plant food plots just to get something green coming out of the ground, but we learned a long time ago that that wasn't good enough. Now we plant our food plots with a desired result and a definite goal in mind.

If we can produce the kind of results we've seen in Morgan County, Georgia, I'm convinced you can do the same thing on you deer lease no matter where you live.

You can make positive changes and see results. However, you've got to follow the steps that we've outlined in this book. You've got to plant the food plots and you've got to put out minerals and supplemental feed. You've got to shoot the does and you've got to let your young bucks go.

Don't give up, and don't get discouraged. It's a long-term program that requires a lot of commitment, but it is possible to see positive results in a relatively short period of time. I know that some hunters are going to read this book and insist that the only reason we've been able to accomplish the things we have on the Banks Farm is because we are blessed to have extremely good whitetail genetics in Morgan County, Georgia.

If you live in an area where whitetail genetics are not the best in the world, don't use that as an excuse. Even with the poorest genetics, you can make a huge difference in the size of your bucks by following the game plan we've outlined in this book. It may take a few years or even a few generations of deer to start turning some of your bucks around. By following the simple steps we've outlined and offering your deer a high-protein diet 12 months a year, you'll see changes in antler mass and size. Aim high, and stick with the program. If you do, I guarantee you that you'll be seeing some power bucks of your own!

Summary — Chapter 10

- Choose your club members carefully.
- Make sure your club members are compatible.
- Choose members that will contribute "sweat equity" – not just money.
- Make it fun, make it a family affair.
- Establish a minimum gross B&C score for trophy bucks.
- A one-buck limit is essential for any trophy management program.
- Keep traffic to a minimum.
- Communication is so important.
- Don't get discouraged.
- Even if you have relatively poor genetics in your area, you can grow bucks with larger antlers and heavier mass if you follow the simple steps in our program. It may take several generations, but you can do it.
- Aim high – be the best you can be!

Club member Carl Wilson took this very unusual 5x2 buck with a bow during the 2001 season. Despite the lop-sided rack, Carl's buck still grossed 130. The buck actually had a cracked skull plate near one antler base that was infected. Chances are, the buck might not have survived this injury.

Chapter 11

Seeing Big Results – Sweat Equity and Making the Dream a Reality

We're very proud of the fact that since 1993, we've produced at least one buck each year on the Banks Farm that has scored right around the 150 mark. We've killed several bucks in the high 140s, and several that gross just over 150. We feel like our success has proven that we've been able to take an average buck with average genetic potential and push him up to the 150-class mark by making sure he has high-protein foods available to him 365 days a year. If we can achieve this through protecting our young bucks, shooting does, and supplemental feeding, we feel like anyone in the Southeast can do the same thing. As an added bonus to any serious trophy management program, you'll always have exceptions that go beyond the normal range. If you're consistently producing 150-class bucks, every once in a while you'll have a buck that goes beyond that point. We've had several in the 160s. In 2001, we finally broke that magical 170-point barrier with a Boone and Crockett qualifier that netted 172 3/8 typical points.

When Dr. Grant Woods visited our property during the summer of 2001, he was amazed to see so much natural deer browse out in the woods. He noted that our deer were not eating any of the new growth on ragweed, poke salad or several other favored deer foods. He told us that these were definite signs that our deer population was down where it needed to be because of the excellent job we had done with shooting our does every year. Every hunter wants to shoot a big buck, but in order to do that, you've got to harvest the does first and let the younger buck go.

If you see an increase in your average deer weights, you're making progress. As mentioned before, we've seen not only a substantial increase in antler size, but we've always seen a noticeable increase in antler mass. We feel like this increase is a direct result of a high-protein diet along with a combination of all of the other management techniques we were using. What other reason could there be?

In the early 1990s when we were just starting out, we had a two-buck limit like the majority of hunting clubs across the South. It never worked. When you know you have the luxury of shooting a second buck, it's easy to take a marginal buck the first time one comes by your stand. That's what was happening with our club. Some of our guys were getting a little anxious, and they'd shoot the first "decent" buck they saw. That's not good enough for a true trophy management program. No matter how much land you have, if every hunter in the club kills two bucks every season, you're putting a huge dent in your buck population.

Say you have 15 members in your "trophy" club and you have a 500-acre lease. Say you have a two-buck limit. Typically, each one of your hunters will claim that he wants to shoot a trophy buck, but in reality, most will be willing to settle for the first 120-class eight-pointer that comes along. With that kind of program, you may lose 10 to 12 bucks or more off your lease each year. Of those 10 to 12 bucks, probably only one or two of them should have been taken. The other ten should have been passed up. If you ever want to produce a real power buck on your lease, you've got to be willing to make some sacrifices. Practicing restraint is one of the biggest sacrifices you'll ever make.

My dad realized the problem with a two-buck limit right away, and he decided to cut it off at the pass. Implementing a one-buck limit is the best thing we ever did on the Banks Farm. People become a lot more selective when they know they only have one chance each season. Not only did our guys become more selective, but they also began to consistently pass up larger and larger bucks. In the past, they might have passed up that 120-class eight-pointer mentioned above with no difficultly. Within a few short years, though, they were beginning to pass up really nice 130-class 10-pointers. Instead of shooting them with a rifle, they were shooting them with their video cameras.

Whenever one of our hunters came back to the cabin and showed everyone else some spectacular video footage of a beautiful 10-pointer that might gross 135 B&C points, it became much easier for the other guys to pass up a buck of that quality the next time they saw one in the woods. More importantly, everyone began to *want* to pass up these bucks because they knew it was a necessary part of the program. They knew in their hearts that the very next year, that same 135-class 10-pointer might be a 150-class power buck. It's a matter of attitude, and that's what you have to do if you ever want to shoot a

Quality buck standards are a must in any serious trophy club. So is a one-buck limit. Here, Pat McDevitt poses with a true quality nine-pointer he shot in 1995. As of that time, it was his best buck to date. Pat's buck netted in the low 130s.

true trophy buck. You can't be willing to settle for second best.

Teamwork Really Pays Off

A lot of hunting clubs never get past the point of being able to pass up bucks in the 130-class range. Hunters come up with all kinds of excuses. I think I've heard most of them. "I thought he was much bigger than he actually was." "I thought his tines were a lot longer." "I saw the buck's main beam and I thought it was a 10-inch G-2 so I shot him. Then I realized his tines were only five-inches long." We all make genuine mistakes from time to time and that's understandable. But if you ever expect to have a true trophy club, every hunter in the club has to grow beyond the excuse level.

Every hunter in your club also has to work together as a team and be willing to put in the sweat equity. You can no longer think in terms of, "I've paid my membership dues so I'm going to get my money's worth by killing *a* buck." Instead, you've got to have the mindset that, "We know there's a 150-class buck out there because he's been seen on our property several times. Eventually, *someone* in the club is going to get a crack at him." And when that someone does kill this particular buck, everyone else in the club should be as happy for that lucky hunter as if they had killed the buck themselves. It's got to be a team effort all the way.

Quality Buck Standards Are a Must

Today the rules of our club state that you can shoot one mountable buck per year, and that buck has to gross 130 B&C points or better, period. Every hunter in our club knows what a 130-class buck looks like. We've made some mistakes in the past, and I'm sure we'll continue to make them occasionally in the future. However, most of our guys are looking for something much larger than a 130-class buck. Therefore, the mistakes are few and far between.

Some clubs have a four-point rule (the buck has to have a minimum of four visible tines on one or both sides). Other clubs have a minimum outside spread rule of 15- to 16-inches. In my own state of Georgia, the Department of Natural Resources began implementing a quality buck program is several South Georgia counties about 10 years ago. Since that time, the program has spread to a number of other Central and South Georgia counties. In some of these quality buck counties, the four-point rule is in effect (that is, four antler points on either side that measure a minimum of one-inch in length). In other counties, a buck has to have a minimum outside spread of 15 inches.

This has been a great boost for Georgia's deer program, and it has done a lot to protect 1 1/2 year old bucks. Like almost all of the eastern states, around 90 percent of the entire buck harvest in Georgia back in the late '80s and early '90s consisted of bucks that were 1 1/2 years old. In many areas, there were almost no older-age class bucks because hunters were killing nearly all of the yearlings. Georgia's pioneer quality buck program has been a great start toward opening the door for even bigger and better management policies in the future. I'd like to see it go even further in the future. There are some counties in Central Georgia that have incredible trophy potential – *if* the bucks could get some age on them.

On the Banks Farm, we wanted to take our trophy program to a much higher level than what the state was doing and we did. As I've mentioned previously, I never liked the idea of the four-point rule or a minimum outside spread rule because it leaves too much to interpretation. For our club, it seemed much simpler to tie our minimum trophy requirements to a gross B&C score. And with a minimum of 130 gross B&C points, a hunter can still take a good buck with his bow and be close to the Pope and Young minimum of 125 net points as well.

In 1997, Carl Wilson shot a nine-pointer that grossed right at 150 B&C points. Carl's buck had a 14-inch inside spread. In 2001, my dad shot a main-framed 4x4 (with a sticker) that grossed 151 B&C points. Dad's buck only had a 14 6/8 inch inside spread. Neither of these bucks would have qualified as legal bucks if there had been a 15-inch inside spread rule in effect, but both bucks were outstanding trophies. Dad's buck had nearly 13-inch tines!

Every rack is different; every buck is unique to itself. If you don't know how to score a deer on the hoof, there are many ways to learn. Studying the pictures in magazines and videos and watching live deer out in the food plots are two of the best ways I know of to learn how to estimate the size of a buck's rack. In many ways, it's just an educated guess until you get him on the ground, anyway, but it'll definitely make you a better hunter if you make the effort to learn something about judging antlers on the hoof.

I've mentioned several times that up until the 2001 season when I killed the Horn Donkey, I hadn't shot an antlered buck on our farm since 1996. My '96 buck grossed just under 150 B&C points and he was definitely a shooter. Between '96 and '01, I let a lot of bucks go. I saw some 130 and 140-class bucks that I did not shoot. I did my part when it came to shooting my share of does. I plowed and planted our food plots just like everyone else in the club. I had no regrets during that five-year period because the guys I hunt with were all passing up 130- to 140-class bucks just like I was. In fact, some of the

Three generations of Banks Farm hunters show off a true "power" buck! My dad Lamar Banks (left) shot this massive main-framed 4x4 during the 2001 season. The big buck grossed 151 B&C points. Looking on with me is my younger brother Loy.

It takes 100 percent to make the program work. In addition to shooting does and protecting younger bucks, you must have a three-pronged nutritional program – planting high-protein food plots, putting out supplemental feed and putting out minerals. With this combination, you can't help but grow some outrageous "power" bucks!

guys shot some video footage of some very impressive deer that were never killed. That's what you have to do if you expect to take things to a higher level and produce some true power bucks.

It takes both time and a lot of commitment, and you have to make some sacrifices along the way. But the returns can be amazing. Once you get all the ingredients stirring around in that pot and start cooking those ingredients, you're going to see results. In fact, if you follow the five keys outlined in Chapter 2, you'll begin to see results inside of two years. Once you start shooting does and protecting your 1 1/2 year old bucks, those bucks will be 3 1/2 years old in two years. Any 2 1/2 year old bucks you had on your place will be 4 1/2 years old in two years. In 24 months, you'll have larger bucks on your lease from the age factor alone. But if you have the vision to start a high-protein planting and supplemental feeding program as well, you'll be well on your way to growing some real power bucks.

I always knew that my day would come sooner or later. It happened on opening day of the 2001 season. And when it did happen, it went beyond my wildest dreams. Me – Jeff Banks – a simple country boy from Morgan County, Georgia, actually shot a buck that qualified for the Boone and Crockett Record Book. It's a dream-come-true and I hope I never wake up from that dream.

In some ways I have to admit that I never really expected to shoot a buck that big. But I always knew it *could* happen on the Banks Farm. I had confidence in that fact. What's more, it easily could have happened to any of the other guys who were hunting on the farm that day.

It Takes 100 Percent to Make the Program Work

If you don't get any other information out of this book that you find helpful, it's important to understand one thing. This may be the most important thing in the entire book. If you're interested in establishing a serious trophy management program with your hunting club, and if you're truly interested in growing some real power bucks in the future, you can't have a partial program and expect it to work. *All* of the ingredients have to be in force before the program will work to its full potential. Once those ingredients are in place, it's also going to take some sweat equity to keep things running smoothly. With all of the guys I hunt with, investing a little sweat equity is a labor of love. It's never work. It's fun.

The essential ingredients I'm talking about are the five keys mentioned in Chapter 2 and elsewhere in the book. It's worth listing them here one last time because there is no way that I can adequately express their importance. They are:

1) **shooting does**
2) **letting smaller bucks go**
3) **planting food plots**
4) **supplemental feeding**
5) **putting out minerals**

Georgia Outdoor News editor Daryl Kirby congratulates me on opening day of the 2001 rifle season in Georgia. It was a big day for all of us at the Banks Farm. We had just done the impossible. We had taken a Boone and Crockett buck!

All of these ingredients work hand-in-hand. All of them have to be in place together like a giant jigsaw puzzle before your trophy management program can reach its full potential. Leave one of piece of the puzzle out, and that puzzle will never be complete. The entire program may fail. You may make some forward progress, but you'll never win the jackpot.

We decided to name this book ***Producing Power Bucks – the Banks Farm Way*** because through trial and error over the past 12 years, we came up with a very simple but successful method for growing trophy bucks on our property. The concept of quality deer management has been around for a long time – at least 25 years. You

Producing Power Bucks co-author Duncan Dobie and I are standing in a field of yuchi clover. Fields like this provide a protein factory for our deer.

may say that we're not doing anything different on the Banks Farm that Al Brothers and Murphy Ray didn't write about in their landmark book **Producing Quality Whitetails** published in 1975. You may insist that we're not doing anything any different from the way the Quality Deer Management Association recommends it be done to its members and to deer hunters everywhere. It's true that we've been practicing quality deer management for a long time on the Banks Farm. But the big difference in our plan is in nutrition, specifically high-protein nutrition. The "Banks Farm way" has taken the nutrition portion of the well-known age, genetics and nutrition formula to a higher level.

 That's all we've done. I like to think of myself as a simple deer hunter. I like to keep things on the farm real simple. Nothing we do is that complicated. But we are willing to give 110 percent to our program by putting in the necessary sweat equity. We do the right things all year long to give our deer all the protein they need. If you'll do the same thing on your property or club lease, you'll produce some real power bucks just like we've been able to do. I guarantee it!

Summary — Chapter 11

- Teamwork really pays off.
- Quality buck standards are a must.
- Learn to score antlers by the B&C measuring system.
- It takes 100 percent to make the program work properly.
- You can do it, too!

The Banks Farm has come a long way in 12 years. In my opinion, taking a true power buck like the Horn Donkey is living proof that high-protein feedings can make a huge difference in your trophy management program. Look at the mass on this buck's antlers!

Chapter 12

The Legacy of the Horn Donkey

Deer hunters are a peculiar breed. We come up with some of the craziest names for big bucks. In my home state of Georgia alone I've heard names like Old Whitey, Big Red, the Phantom, and the Silver Bullet Buck given to legendary bucks. When my good friend Charles Hendrix came up with the name "Horn Donkey" for the buck I would end up taking on opening day of the 2001 season, it was one of those spontaneous things that seemed to fit the big buck to a "t". So the name stuck.

It's not often that a country boy like me gets to shoot a super power buck like the Horn Donkey. By now, you may be getting tired of hearing his name since I've been talking about him throughout this book. I hope you'll bear with me a little longer. In truth, the Horn Donkey is probably the driving force behind the writing of this book. He's certainly been a big event in my life, and also in the lives of everyone who hunts on the Banks Farm. So I hope you can understand some of our excitement.

My dad has been a serious deer hunter all his life. I guess it was only natural that some of his enthusiasm for giant bucks would rub-off on me. I told the story in Chapter 2 about how Dad had the Buck Ashe trophy hanging in our house for almost a year during the late '60s. I also told the story about seeing the Tom Cooper "Silver Bullet" buck taken at B.F. Grant WMA in 1974. I was only four or five years old when the Buck Ashe trophy hung in our house. Because of Dad's excitement about that trophy, it made a lasting impression on me. One of the reasons it made such a huge impression on me was because Dad always told us it was the biggest whitetail ever taken in Georgia.

I was 10 when I saw the Tom Cooper buck. I think it made an even bigger impression on me because by then I was old enough to understand what a rare privilege it was just to see a buck like that. Furthermore, the knowledge that a buck of that size had been killed only a few miles from where we lived at the time got us all fired up. (As mentioned in Chapter 2, the Tom Cooper buck scored 215 7/8 non-typical B&C points and went on to win the National Rifle Association's Silver Bullet Award for 1974.)

Now, looking back, it's almost like seeing those two giant bucks at an early age had something to do with preparing me for the future. Little did I know back then that I would some day kill a record-book buck of my own.

All I can tell you is that I grew up with a passion for deer hunting. Like most hunters, I killed my share of small bucks in the beginning. But when I finally began to mature as a hunter, I suddenly wanted to shoot bigger and better deer. I wanted to shoot a real trophy whitetail like the Tom Cooper buck. I had seen pictures and read stories about some of Georgia's biggest bucks ever taken, and I knew that we lived in an area where shooting a big buck was entirely possible. In fact, when Duncan Dobie's book **Georgia's Greatest Whitetails** came out in 1986, I read every story in it twice and I bought extra copies to give all my friends. Little did I know back then that we

The Legacy of the Horn Donkey

During the summer of 2001, we got some trail camera photos of a really nice buck in velvet at the feeder in our front yard. This photo, taken in late August, was the first of at least eight different photos we would get of the buck that would later be named the "Horn Donkey."

Even after he shed his velvet, we didn't realize the Horn Donkey was as big as he turned out to be. In this photo, you can see how he dwarfs the young eight-pointer. We knew he was a fine trophy, but we never dreamed that his rack would gross 197 B&C points (including non-typical points).

would one day be writing a book together with *my* story in it!

I knew that shooting a big buck was possible, but I knew it probably wouldn't happen by chance alone. In 1990, I realized the only way we could "make it happen" was if we started a serious trophy management program.

Twelve years later we had come a long way with our program. If we had never seen a buck like the Horn Donkey on our farm, I'd still be very proud of how far we'd come and how much we'd learned during all those years. As things turned, however, the giant buck that later earned such an unusual nickname just happened to show up at a feeder during the late summer of 2001. I don't mind telling you that it changed all our lives forever.

In 1995 shortly after building our new house and moving to the farm, I had built a feeder in my front yard so that we could watch deer from the house. My wife Michelle and I only had one child at the time, a daughter, Kimberly, but we soon were blessed with our son "J.L." as well. As the kids grew older, I thought it would be a lot of fun to be able to watch deer from the house. I also have a large food plot in my front yard. It sure beats cutting grass.

In the summer of 2001, we put a trail camera near the feeder in hopes of getting some good night pictures. We weren't disappointed. We got a good picture of a really nice buck one night at the feeder. He was still in velvet. I had no idea how big he was at the time. I just thought he was a nice, healthy buck. In fact, we didn't really pay much attention to that first picture.

A short time later, Charles Hendrix got some trail camera pictures of the same buck across the road on his property. Charles immediately christened the big buck the *Horn Donkey*. All the guys in our club started studying the pictures. Everyone agreed he was a good mature

If you want to say that I got lucky on opening day of the 2001 rifle season in Georgia, I'll have to agree. My definition of luck is where opportunity and preparation meet! Here, my son J.L. and I show off the Horn Donkey's massive 16-point rack. I wonder what he'd been eating for the last 5 1/2 years.

buck, but we still had no idea just how *good* he was. If someone had told me then that they thought the buck in those photos was a record-book contender, I would have laughed in their face. If someone had insisted that he would gross in the mid 190s, I would have laughed even louder.

In the past, some of our trail camera photos had made certain bucks look much bigger than they actually were. In the case of the Horn Donkey, just the opposite took place. The pictures made him look smaller than he actually was. For one thing, he appeared to have a very high rack with long tines. Because of this, his rack didn't appear to have unusually heavy mass. We would later find out after he was on the ground that his bases alone measured 6 3/8 inches and 6 2/8 inches. In other words, he had incredibly heavy mass. (Somehow, that buck must have been getting some serious protein in his diet. I wonder where it came from.)

It was also hard to judge the Horn Donkey's weight from the photos. As we would later find out, he carried a live weight of 232 pounds. But you couldn't tell that from the photos, and his body size also contributed toward making his antlers appear smaller. At any rate, we all tried to score him by the pictures as mentioned in Chapter 1. A couple of the guys thought he might score around 150. I had him in the high 160s. My cousin Danny Keever was the closest. He scored him at around 180. Danny said, "Man, he can't be that big, can he?"

We had no idea what the Horn Donkey would actually score. We were just guessing. Our estimates were all

After getting at least eight different photos of the Horn Donkey in a number of different camera locations, we started seeing a definite pattern to the big buck's movements. He was probably running a long, narrow circle. We had about one-half of the circle figured out. This knowledge had a lot to do with where I decided to hunt on opening day. It paid off big time!

gross scores. That's the way we always score our bucks on the Banks Farm. If a buck has a drop tine or several sticker points, we add those into the gross score.

To our knowledge, no one had ever seen this buck for certain in the past. That's not unusual on the Banks Farm. Like all hunting clubs, we've gotten video footage and trail cameras photos of certain bucks that no one had ever seen before. No one ever saw them again, either. They just drifted in and drifted out like big bucks often do. However, we do seem to be able to hold a lot of the big bucks on our farm because we often see the same bucks over and over again. And even though no one could say for sure that they had seen the Horn Donkey on the farm, it's almost a certainty that someone passed him up when he was younger.

The only one of our hunters who might have seen the Horn Donkey in the flesh was Carl Wilson. The year before during the 2000 season, Carl had watched two mature bucks with large, high racks run out of the woods into a field chasing a doe. Both bucks were grunting, and both bucks had impressive antlers. Carl thinks it's possible that one of those bucks might have been him. However, no one ever saw those two again that season.

On October 25, only three days before rifle season opened, another excellent photo of the Horn Donkey was taken at night with one of our trail cameras. By then we knew he was an exceptional trophy buck. We were just beginning to realize that he might be bigger than we had originally thought, and we knew he would probably gross at least 160 B&C points. By opening day of rifle season, everyone in the club was well aware of him. The Horn Donkey was fair game for anyone who hunted on the Banks Farm.

An Unforgettable Opening Day

In addition to showing us that the Horn Donkey had an exceptional trophy rack, the various photos taken with different trail cameras in different locations were also beginning to show a definite pattern to his movements. From studying the photos, I had a good picture in my mind of how the buck was moving through our property. Although I hunted quite a bit during archery season, I purposely stayed out of the area he seemed to be using.

From the various locations of where the photos had been taken, we figured he was running a typical 3/4 mile loop. One part of that loop came through the area where I intended to hunt. I decided early in the season that a buck as big as the Horn Donkey should be hunted during the peak of the rut. In our part of Georgia, the rut usually peaks between November 5 and November 15. You can pretty much count on that 10-day period every year. I knew exactly where I wanted to hunt, but I decided to stay out of that area until the rut was in full swing in early November.

On October 27, opening day of gun season, however, conditions seemed perfect. I decided to slip into the area and try my luck from a permanent platform stand we had built several years earlier. (In case I haven't mentioned this before, my definition of luck is where preparation and opportunity meet.) It was a little sooner than I had planned, but I guess I just couldn't wait any longer.

The weather was just right. It was a cool 35 degrees that morning. It was a great day to be in the woods. I started seeing deer right away. I saw several young bucks. My stand was located in a long hardwood bottom containing mature timber. It was surrounded by pine plantations on two sides. A good size food plot planted in ladino clover stood nearby. The stand of hardwoods

Here's another shot of me posed with my good friend Daryl Kirby. When this picture was taken, I was floating on clouds!

around me contained a lot of oak trees, and there were a lot of acorns on the ground. The deer were really eating those acorns.

Everybody knew the Horn Donkey was around but nobody knew exactly where he might show up. He was fair game for anyone who was hunting that day and each one of us was hoping to see him. I just happened to be in the right place at the right time.

Billy Young was hunting from a stand not to far from me overlooking what we call the wind row field. He was also surrounded by pine plantation and he was in a very good location. Ordinarily, the field he was hunting over would have been full of waist-high oats and yuchi clover. But because of the drought that year, the wind row field was way behind in production. Had it been full of good food like it should have been, I honestly believe the Horn Donkey would probably have come out in front of Billy. Judging by the lay of the land and the way we know deer move through that area on a regular basis, that would have been the Horn Donkey's most predictable route. But for some unknown reason, he didn't. He came out in front of me instead.

Even though it was opening day of rifle season, I had my bow with me just in case. We had all been doing a lot of bow hunting earlier in the season, and I had been hoping to shoot a good buck with my bow. If the opportunity presented itself on opening day, that's still what I planned on doing. I also had my rifle and my video camera. The morning hours passed without incident.

After lunch, I returned to my stand early in the afternoon. I had a good book with me and I did some reading to pass the time. I know this sounds crazy, but I love to read when I'm sitting in a stand for a several hours. It's so peaceful in the woods. It's hard for me to find time to read in other places. Even though I'm reading, I still manage to stay very alert.

It was a beautiful afternoon to be in the woods – just like the morning had been. I can't begin to tell you how much I was enjoying the day. About 5:30 p.m., a doe came off the hill to my left out of the planted pines. She started working her way toward me. A few minutes later I saw another doe farther away behind the first one. They were both eating acorns from the nearby white oaks and water oaks. Five or ten minutes went by and the doe closest to me suddenly threw her head up and looked back.

If you're a serious deer hunter, you know the kind of look I'm talking about. It can only mean one of two things. 1) Someone is coming through the woods and the deer is getting ready to spook, or 2) a buck is coming.

This time, I just had a feeling that it was a buck. I said to myself, "Man, where art thou?"

Then I looked up and saw a buck coming down the hill toward me. He had come out of the same pine plantation and he was headed right for the doe feeding closest to me. He was totally focused on that doe. He wasn't interested in feeding or doing anything else.

The only reason he came out of those pines was because of that doe. Furthermore, that's the only reason I got him that day.

The minute I saw him, I knew it was the Horn Donkey. I said to myself, "'Man, here comes the Horn Donkey and that joker's a real hoss!"

He looked to be every bit of a 160- to 170-point buck. He was still 85- to 90-yards away and all I could see were parts of his rack as he walked through the woods. He came all the way down the hill toward the doe through a natural funnel where, in the past, I'd seen a lot of nice bucks come through.

When he got to the bottom of the hill he stopped. It was clear that he was trying to tend the doe closest to me. Typical of the way does often act, though, she wasn't paying any attention to him. She was just eating acorns in the hardwood bottom. As she came around to my right, he looked at her and made a very loud grunt. It sounded like the grunt of a very mature buck. It sent a chill down my back. I was still holding my bow.

Actually, I was so transfixed just watching him that I never even thought about picking up my rifle. I was hoping and praying that he might step within bow range. If he did, I knew he would be the trophy of a lifetime with a bow. I love to bow hunt. I've taken many, many does with my bow, but I've never really killed a big buck with a bow and I've always wanted to. Someday I'm going to make it happen.

Don't get me wrong. I was thrilled to shoot him with a rifle. But during the time I was holding that bow I can't begin to tell you how it made me feel. It was an incredible sensation. There are some things money can't buy. The six or eight minutes I watched that buck in front of me was something I'll never forget. It was exciting, unbelievable and awesome all at the same time. It was one of the most amazing experiences I've ever had in the woods.

All of a sudden the doe went up the hill and squatted to urinate. When she was finished, she kind of hopped back down the hill to where she had been feeding. He immediately walked over to where she had squatted and started sniffing the ground. Then he threw his head back and started lip curling. I had seen pictures of bucks lip-curling in books, but I'd never seen it in real life. It was awesome to watch. It was almost like I was watching a video. What's more, when those antlers went back, I could really see the mass of his rack and the length of his tines. I said to myself, "Man, he's much bigger than I thought. He's really something! He might even go well beyond 170!"

Then he came down the hill doing a fast, shuffling walk like I'd seen rutting bucks do many times before. He turned around and faced her. He got around on her right side and grunted one more time like he was trying to herd her. The second grunt was not as loud as the first but you could have easily heard it in that bottom 100 yards away. He wasn't paying any attention to the other doe that was back behind the first doe.

The closest doe ran ahead to my right and stopped. I put my range-finder on her and it registered 23 yards. He stayed right where he was – pawing the ground and banging his antlers in some limbs above the privet. My rangefinder had him at 44 yards but I couldn't shoot

A buck like this deserves to be mounted by a top-notch whitetail taxidermist. Our good friend Ricky Smith of Newborn, Georgia, did a jam-up job on the Horn Donkey.

because it was too thick. I was hoping he was going to follow the doe. I was all ready to try for a shot with my bow when the other doe started moving across the bottom. He suddenly turned and started quartering back toward her.

In 1994, when I shot my biggest buck ever as of that time – a buck that grossed 161 2/8 points – that buck did the same thing. He left the doe he was with and started going toward another doe. Because of that second doe I almost lost my chance for a shot. I was afraid the same thing might happen now. If it did, the Horn Donkey would be going away from me through some really thick privet.

So I had to make a quick decision. I carefully hung my bow up on the hook, and grabbed my rifle – a Weatherby .300 Magnum. I put my gun across my lap, and set the video camera down between my feet and turned it on. I was hoping to get some good video footage of him but it didn't work out. He turned again like he might try to walk away and I made the decision to shoot him with my rifle. I quickly got him in my scope and squeezed the trigger. The range was 44 yards.

He ran a short distance and collapsed. The first thing I did was call Carl Wilson on my cell phone to tell him the news. Carl started calling everyone else and word spread fast. I also called Charles Hendrix who was hunting on his property next door. Charles said, "Don't do anything until I get there. I'm going to bring my tape and measure him."

Charles started green-scoring the big buck right where he had fallen. We came up with a gross score of 191 points. That was very close to the official gross typical score. Officially, the Horn Donkey's rack actually grossed 189 2/8 points. With the non-typical points added in, he grossed 197 3/8 points. It was unbelievable!

Charles said, "Let's start dragging him out of here."

I said, "I'm not dragging him anywhere. I'm leaving him right where he is until everyone in the club gets here. We're all going to drag him out together. Everyone's earned that right! Anyway, how many times does a deer hunter get to drag a Boone and Crockett buck out of the woods, especially one that grosses nearly 200 points?"

Charles understood exactly where I was coming from. After 12 years of hard work, we had actually achieved our goal of producing a buck on the Banks Farm that would qualify for the all-time Boone and Crockett Record Book. It's important to emphasize the "we" part in the equation. Even though I happened to be the lucky hunter who had pulled the trigger, it had been a team effort all the way. Just like a football team that wins the Super Bowl, every single member of our club had had something to do with growing the Horn Donkey.

We left the big buck right where he had fallen. Within the hour, everyone who had been out hunting that day, plus a lot of other people including most of my family – Michelle, J.L., Kimberly and my mom and dad – had arrived at the scene to share the excitement and see in person the giant Morgan County buck known as the Horn Donkey. Even the Morgan County game warden, Mr. Thomas Bernard, was there to share in the excitement. We took dozens of photos and got good video footage dragging him out of the woods.

The Horn Donkey was found to be 5 1/2 years old. His live-weight tipped the scales at 232-pounds. He field-dressed at 195-pounds. After the 60-day drying period, Mr. Bill Cooper, who in my opinion is one of Georgia's best B&C scorers, officially measured the giant 16-point rack at 172 3/8 typical B&C points.

Once again, I want to say that I just happened to be in the right place at the right time that day. Anyone who was hunting that day could have gotten him just as easily as I did. I consider myself a die-hard trophy hunter and I'd love to tell you that it was my great skill as a whitetail hunter that brought down the Horn Donkey. But it wasn't. Countless hours of sweat equity by a lot of dedicated people went into the taking of that buck. Patience and determination also played a key role in attaining our "impossible" dream. I firmly believe that our high-protein food plots and all of the other aspects of our trophy management program had everything to do with growing a buck the size of the Horn Donkey. What's more, I also believe that we'll see more power bucks like the Horn Donkey on the Banks Farm in the future.

Whenever I think of the Horn Donkey, I can't help but think about our rich fields filled with yuchi clover and oats. I think about our iron and clay peas, and I think about our alfalfa. I think about our feeders and all the minerals we put out. I think about all the controlled burning we do to create new browse. I think about all the does we've taken during the last 12 years and all the nice bucks our guys have passed up. I think about the 10 or 12 dedicated guys that I hunt with whose hard work made it all possible. Then I thank the Good Lord for being one deer hunter who has truly been blessed.

If we can do the impossible in Morgan County, Georgia, I'm convinced that you can do it too, no matter where you hunt. I've said it before. I'm just a simple deer hunter. Nothing we've done on the Banks Farm is all that complicated. No matter where you hunt in the Southeastern United States, if you'll follow some of the simple guidelines that I've discussed in this book for planting high protein food plots and for putting out supplemental feed and minerals, you, too, can grow some power bucks of your own!

Score Sheet

Hunter: Jeff Banks
Location: Morgan County, Georgia
Date Killed: October 27, 2001

The Horn Donkey

Scorable Points 16	Abnormal points:		
Tip to Tip Spread 9 1/8	1R – 1 3/8		
Greatest Spread 19 4/8	3L – 6 6/8		
Inside Spread 17	Total 8 1/8		
Areas measured	**Right**	**Left**	**Difference**
Main Beam	25 6/8	26 3/8	5/8
1st point (G-1)	7 5/8	6 2/8	1 3/8
2nd point (G-2)	10 1/8	10 2/8	1/8
3rd point (G-3)	9 7/8	12 1/8	2 2/8
4th point (G-4)	6 4/8	7 3/8	7/8
5th point (G-5)	2 1/8	4 3/8	2 2/8
(H-1)	6 3/8	6 2/8	1/8
(H-2)	5	5 2/8	2/8
(H-3)	5	5 6/8	6/8
(H-4)	4 7/8	5	1/8
Totals	83 2/8	89	16 7/8

*Gross typical score 189 2/8
Total asymmetry deductions 8 6/8
Net typical score..180 4/8
Total abnormal points 8 1/8
Final typical score 172 3/8

*(Note: The Horn Donkey's gross typical score including abnormal points is 197 3/8.)

After the required 60 day drying period, Bill Cooper of Tifton, Georgia, officially measured the Horn Donkey's rack at 172 3/8 typical B&C points. Bill is not only a well-known biologist with the Georgia DNR, he is also one of the top B&C scorers in the nation.

Chapter 13

Banks Farm Checklist

January -
- Begin supplemental feeding for deer (or as soon as deer season is over).
- Build new feeders; repair old feeders if necessary.
- Begin controlled burning (January, February and March).

February -
- Continue supplemental feeding for deer.
- Put out minerals for deer (once yearly).
- Continue controlled burning.
- Get soil tests for summer food plots.
- Fertilize oak trees and honeysuckle around fields.
- Plant new trees and shrubs for deer and turkeys as necessary (oak trees, honeysuckle, persimmon, dogwood, and crabapple).

March -
- Continue supplemental feeding for deer.
- Continue controlled burning.
- Begin spring turkey season.
- Look for shed antlers in woods.

April -
- Continue supplemental feeding for deer.
- Prepare soil for planting iron and clay peas.
- Harrow fields.
- Check mineral licks for heavy use.
- Continue hunting spring gobblers.
- Continue looking for shed antlers.

May -
- Continue supplemental feeding for deer.
- Check mineral licks for heavy use.
- Continue hunting spring gobblers.
- If necessary, turn under some food plots containing yuchi clover and oats and replant in iron and clay peas for continuous rotation of high-protein feed. (Note: Obviously, if you have an outstanding crop of standing yuchi and oats, you'll want to leave it alone.)
- Plant iron and clay peas.
- Clip clover and alfalfa fields to help control weeds and to stimulate new growth.
- Plant dove fields (millet and sunflowers).

June -
- Continue supplemental feeding for deer.
- Check mineral licks for heavy use.
- Put out trail cameras and check weekly.

July -
- Continue supplemental feeding for deer.
- Begin practicing for bow season.
- Get soil tests for fall food plots.
- Lime all food plots as needed.
- Check trail cameras.
- Check minerals for heavy use.
- Trim limbs around fields and roads.
- Replant peas if necessary.
- Begin to watch and video deer in food plots. Study big bucks.

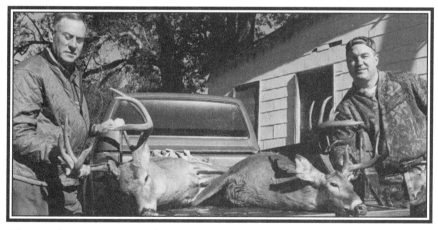

The "Banks Farm Way" produced this truckload of power bucks taken in 1999. On the left is my dad Lamar Banks with a thick-antlered eight-pointer that grossed 131 points. Dad's buck weighed in at 205 pounds (live weight). On the right, Pat McDevitt shows off a beautiful 5 x 5 with stickers that we dubbed "Little Monster." He grossed 141 B&C points. You can grow bick bucks on your land too, if you follow the simple management steps outlined in this book.

August -
- Continue supplemental feeding for deer.
- Bush-hog farm roads and remove any fallen trees.
- Continue practicing for bow season.
- Check trail cameras
- Continue videoing big bucks in food plots.
- Check all permanent and tower stands for any needed repairs.
- Sight-in rifles for gun season.

September -
- Make sure all feeders are empty before bow season opens.
- Plant yuchi arrowleaf clover and oats.
- Plant alfalfa grazing.
- Plant ladino clover.
- Continue watching food plots and videoing big bucks.
- Begin archery season.

October -
- Continue archery season.
- Begin rifle season.
- Fine-tune rifle accuracy.

November -
- Continue rifle season.
- Begin shooting does after Thanksgiving.
- Find out about Hunters for the Hungry programs.

December -
- Continue shooting does.
- End the season with a fine trophy buck!

My good friend Ryan Klesko proudly shows off his Pope & Young trophy taken on the Banks Farm in '98. Ryan owns several tracts in Georgia and he's really into trophy management.

Valuable Resources for Your Trophy Management Program

- Call your local County Extension Agent for soil tests

- Ducks Unlimited
  1303 Kennedy Drive
  Griffin, Georgia 30224

- Georgia Department of Natural Resources
  2117 U.S. Hwy. 78 SE
  Social Circle, Georgia 30025

- Georgia Forestry Association
  (For help with burning, all states have forestry associations and most counties have regional offices.)

- Georgia Forest Stewardship Program
  1-800-GATREES

- Godfrey Feed and Seed
  P.O. Box 488
  Madison, Georgia 30650
  (706) 342-0264

- Hunters for the Hungry (in Georgia)
  WWW.GOHUNTGEORGIA.COM
  (404) 892-FEED
(Note: other states have similar programs)

- National Wild Turkey Federation
  P.O. Box 530
  Edgefield, South Carolina 29824
  1-800-THE-NWTF

- Pennington Seed, Inc.
  P.O. Box 290
  Madison, Georgia 30650
  (706) 342-1234

- Piedmont Ag Service
  2120 Pierce Dairy Rd.
  Madison, Georgia 30650
  (706) 342-5694

- Quality Deer Management Association
  P.O. Box 227
  Watkinsville, Georgia 30677
  1-800-209-DEER
  www. qdma.com

- For Whitetail Consultants contact:
  Southern Forestry Consultants
  www.soforest.com
  (229) 246-5785

- Grant R. Woods, PHD
  Woods and Associates Inc.
  Wildlife Biologists
  427 DunGannon Dr.
  Abbeville, South Carolina 29620
  (864) 229-6112

- Turn in Poachers (Georgia)
  1-800-241-4113

NOTES

NOTES

NOTES

NOTES